Metamaterials and Their Devices, Second Edition

Metamaterials and Their Devices, Second Edition

Guest Editor

YoungPak Lee

MDPI

Basel • Beijing • Wuhan • Barcelona • Belgrade • Novi Sad • Cluj • Manchester

Guest Editor
YoungPak Lee
Department of Optical
Science and Engineering
Fudan University
Shanghai
China

Editorial Office
MDPI AG
Grosspeteranlage 5
4052 Basel, Switzerland

This is a reprint of the Special Issue, published open access by the journal *Crystals* (ISSN 2073-4352), freely accessible at: https://www.mdpi.com/journal/crystals/special_issues/CFT76J015N.

For citation purposes, cite each article independently as indicated on the article page online and as indicated below:

Lastname, A.A.; Lastname, B.B. Article Title. *Journal Name* **Year**, *Volume Number*, Page Range.

ISBN 978-3-7258-7290-9 (Hbk)
ISBN 978-3-7258-7291-6 (PDF)
https://doi.org/10.3390/books978-3-7258-7291-6

Contents

crystals

MDPI

Editorial

Metamaterials and Their Devices, Second Edition

YoungPak Lee [1,2]

1 Department of Optical Science and Engineering, Fudan University, Shanghai 200433, China; yplee@hanyang.ac.kr or yplee@fudan.edu.cn; Tel.: +82-10-3739-6514
2 Department of Physics, Hanyang University, Seoul 04763, Republic of Korea

Over the past two decades, metamaterials (MMs) have led a revolution in new material science through the artificial arrangement of electric and magnetic resonance structures (meta-atoms) at subwavelength scale. In particular, they have enriched the fundamental rules of matter–light interactions, such as slow light, super resolution, super-lensing, and electromagnetic (EM) cloaking. The main reason for the attention paid to MMs is that they are very close in appearance to real life materials, such as perfect absorbers. EM MMs reveal remarkable responses to the incident EM wave, such as negative-refraction index, extraordinary optical transmission, electromagnetically induced transparency-like effects, and ultra-thin and broadband absorbers. The designed structures, the structural parameters, and the properties of used materials yield the effective electric permittivity (εeff(ω)) and the effective magnetic permeability (μeff(ω)) of overall MMs, based on the effective-medium theory. Studies on the control of EM response and its spatial distribution and dispersion are ripe and lead to potential and almost-realized applications. There are emerging fields in MM research, such as nonlinear, switchable, gain-assisted, sensor, quantum, and coding MMs, all representing a variety of MM applications. This Special Issue on "Metamaterials and their devices, Second Edition" aims to cover a broad range of topics, including MMs, lattice MMs, crystal materials and structures, plasmonic and dielectric MMs, photonic crystals, phononic crystals, metasurfaces, relevant fundamental issues, emerging fields for MMs, electromagnetic response, magnetic resonance, electric resonance, numerical methods, and applications. This Special Issue contains a mixture of review articles and original contributions.

A long-wave infrared circularly polarized photodetector based on an array of trapezoidal silicon pillars was investigated by Cheng et al. [1]. Integrating metasurface-based polarizing filters atop photodetectors enables the expansion of detection capabilities from intensity to polarization, offering significant potential for applications requiring high-precision discrimination in scientific, industrial, and defense sectors. However, such metasurfaces often introduce optical efficiency losses. They present a long-wave infrared (8.6 μm) circularly polarized photodetector capable of direct chiral discrimination, eliminating the need for additional optical components. The polarization selectivity arises from guided-mode resonances excited by two horizontally offset right-trapezoidal unit cells within a chiral metasurface. This design exhibits a pronounced transmittance contrast (~100%) between left circularly polarized light and right circularly polarized light while maintaining fabrication simplicity via a conventional single-step lithographic process. The proposed detector is expected to achieve high-dimensional physical characterization by resolving polarization-encoded vectorial information, demonstrating enhanced performance in complex environments.

Wavelength conversion in photonic crystal fibers via multiple Raman redshifts and soliton spectral tunneling was studied by Deng et al. [2]. The soliton spectral tunneling

https://doi.org/10.3390/cryst16020092

effect in photonic crystal fibers with three zero-dispersion wavelengths is an effective way to realize pulse wavelength conversion. However, due to the limitation of the soliton splitting mechanism, forming a tunneling soliton with higher energy and greater width by increasing the number of Raman redshifts remains a key challenge. Airy prime pulses generate polychromatic Raman solitons with small truncation coefficients, and they converge into a stable soliton after the soliton tunneling effect, which provides a new possibility for addressing this problem. This paper discussed how to control the energy, width, and central wavelength characteristics of the tunneling solitons generated in the photonic crystal fiber with three zero-dispersion wavelengths by adjusting the truncation coefficient α, peak power P, and central wavelength λ of the initial Airy prime pulse. The results showed that the smaller the truncation coefficient α, the greater the number of Raman self-frequency shifts and the greater the energy of the formed tunneling soliton. The increase in initial power P_0 will lead to an increase in tunneling soliton width. The larger initial center wavelength λ significantly increases both the width and the center-wavelength position of the tunneling soliton. These findings provide a theoretical basis for the application of Airy prime pulses in ultrafast optical wavelength control and new light source development.

Line-defect phononic crystal structure for directional enhancement detection of weak acoustic signals was examined by Zhang et al. [3]. Effective detection of acoustic signals plays a crucial role in numerous fields, including industrial equipment fault prediction and environmental monitoring. Acoustic sensing technology, owing to its substantial information carrying capacity and non-contact measurement advantages, has garnered widespread attention in relevant applications. However, the effective detection of weak target acoustic signals amidst strong noise interference remains a critical challenge in this field. The core bottleneck lies in the difficulty of traditional detection methods to simultaneously achieve both high sensitivity and high directionality. To address this limitation, this work proposed a line-defect phononic crystal structure that enabled directional enhancement and detection of weak target signals under intense spatial noise interference by coupling defect-state localization characteristics with anisotropy mechanisms. Through theoretical derivation and finite element numerical simulation, the directional enhancement properties of this structure were systematically validated. Furthermore, numerical simulations were conducted to validate the detection of weak harmonic signals and weak bearing fault signals under strong spatial noise interference. The results demonstrated that this line-defect phononic crystal structure exhibited high feasibility and outstanding performance in detecting weak acoustic signals. This work provides novel insights for developing new acoustic detection methods combining high sensitivity with high directivity, showcasing unique advantages and broad application prospects in acoustic signal sensing, enhancement, and localization.

Tandem neural network based design of acoustic MMs for low-frequency vibration reduction in automobiles was also investigated by Deng et al. [4]. Automotive NVH (noise, vibration, and harshness) performance significantly impacts driving comfort and traffic safety. Vehicles exhibiting superior NVH characteristics are more likely to achieve consumer acceptance and enhance competitiveness in the marketplace. In the development of automotive NVH performance, traditional vibration reduction methods have proven to be mature and widely implemented. However, due to constraints related to size and weight, these methods typically address only high-frequency vibration control. Consequently, they struggle to effectively mitigate vehicle body and component vibration noise at frequencies below 200 Hz. In recent years, acoustic metamaterials (AMMs) have emerged as a promising solution for suppressing low-frequency vibrations. This development offers a novel approach for low-frequency vibration control. Nevertheless, conventional design methodologies for AMMs predominantly rely on empirical knowledge and necessitate continuous parameter adjustments to achieve desired bandgap characteristics—an endeavor

https://doi.org/10.3390/cryst16020092

that entails extensive calculations and considerable time investment. With advancements in machine learning technology, more efficient design strategies have become feasible. This paper presented a tandem neural network specifically developed for the design of AMMs. The trained neural network was capable of deriving both the bandgap characteristics from the design parameters of AMMs as well as deducing requisite design parameters based on specified bandgap targets. Focusing on addressing low-frequency vibrations in the back frame of automobile seats, this method facilitated the determination of necessary AMM design parameters. Experimental results demonstrated that this approach could effectively guide AMM designs with both speed and accuracy, and the designed AMMs achieved an impressive vibration attenuation rate of 63.6%.

Cheng et al. [5] elucidated the design of multifunctional polarization waveplates based on thermal phase-change metasurfaces. The switching function of traditional waveplates necessitates mechanical replacement or the superimposition of multiple waveplates, which gives rise to a complex system and a large volume. They have devised a multifunctional micro-waveplate based on the COMSOL simulation platform (v5.6), which concurrently integrated the compact nature of metasurfaces and the dynamic regulatory features of phase-change materials. When the phase-change material is in the crystalline phase, the metasurface possesses the functionality of a half-waveplate and is capable of performing chirality inversion of circularly polarized light within the wavelength ranges of 1.45 to 1.52 μm and 1.56 to 1.61 μm. When the phase-change material is in the amorphous phase, the metasurface serves as a quarter-waveplate and can achieve the conversion between linear and circular polarization through a 90° phase delay. The phase-change metasurface breaks through the constraint of fixed functions of traditional optical waveplates, facilitating the development of optical systems towards miniaturization, intelligence, and low power consumption, and providing a crucial technical route for the next generation of photonic integration and dynamic optical applications.

Evolution of phonon spectral energy density in superlattice structures was examined by Nasiri and Wang [6]. Superlattices are a distinctive class of artificial nanostructures formed by the periodic stacking of two or more materials. The high density of interfaces in these structures often gives rise to exotic physical properties. In the context of thermal transport, it is well established that such interfaces can significantly scatter particle-like phonons while also inducing constructive or destructive interference in wave-like phonons, depending on the relationship between the phonons' coherence lengths and the superlattice's period thickness. In this work, they systematically investigated the effect of temperature on the spectral energy density of phonon modes in superlattices. Additionally, we examine how variations in superlattice period thickness influence phonon lifetimes and energy density. Our findings provide critical insights into the spectral phonon properties of superlattices, particularly in terms of their coherence and lifetimes.

The design and processing of metamaterials was reviewed by Matei et al. [7]. Metamaterials represent artificially structured materials that exhibit unusual properties, such as a negative refractive index, negative permeability and permittivity, negative Poisson's ratios, and unique optical effects, which are inaccessible in natural materials. According to recent developments, novel devices and tools based on MMs are attracting great interest as they offer improved performance, functionality, sensitivity, biocompatibility, complex structures, and design freedom. Leveraging numerical design approaches, such as finite element analysis and finite difference time domain methods, researchers have tailored metamaterials to meet specific requirements in various areas through a range of manufacturing techniques. These materials can be broadly classified into optical, mechanical, thermal, electromagnetic, and acoustic categories based on their properties and intended use. The choice of fabrication method depends heavily on the specific application,

https://doi.org/10.3390/cryst16020092

the desired scale, and the complexity of the metamaterial design. These manufacturing methods can be broadly divided into top-down and bottom-up approaches, while each of them has advantages and limitations and offers valuable pathways for the development of the final product. This review offered a basic overview of metamaterials, covering their fundamental principles, fabrication and characterization techniques, and current design methodologies. It also explored their diverse applications, including specific case studies in medicine, while addressing existing limitations and challenges. Finally, this review highlights future perspectives, emphasizing the need for continued innovation in fabrication and characterization to unlock the full potential of MMs.

Soroosh et al. [8] studied a compact and fast resonant cavity-based encoder on a photonic crystal platform. The soliton spectral tunneling effect in photonic crystal fibers with three zero-dispersion wavelengths is an effective way to realize pulse wavelength conversion. However, due to the limitation of the soliton splitting mechanism, forming a tunneling soliton with higher energy and greater width by increasing the number of Raman redshifts remains a key challenge. Airy prime pulses generate polychromatic Raman solitons with small truncation coefficients, and they converge into a stable soliton after the soliton tunneling effect, which provides a new possibility for addressing this problem. This paper discussed how to control the energy, width, and central wavelength characteristics of the tunneling solitons generated in the photonic crystal fiber with three zero-dispersion wavelengths by adjusting the truncation coefficient α, peak power P, and central wavelength λ of the initial Airy prime pulse. The results showed that the smaller the truncation coefficient α, the greater the number of Raman self-frequency shifts and the greater the energy of the formed tunneling soliton. The increase in initial power P_0 led to an increase in tunneling soliton width. The larger initial center wavelength λ significantly increased the width and center-wavelength position of the tunneling soliton. These findings provide a theoretical basis for the application of Airy prime pulses in ultrafast optical wavelength control and new light source development.

Sun et al. [9] examined an embedded rough-neck Helmholtz resonator low-frequency acoustic attenuator. Effective detection of acoustic signals plays a crucial role in numerous fields, including industrial equipment fault prediction and environmental monitoring. Acoustic sensing technology, owing to its substantial information-carrying capacity and non-contact measurement advantages, has garnered widespread attention in relevant applications. However, the effective detection of weak target acoustic signals amidst strong noise interference remains a critical challenge in this field. The core bottleneck lies in the difficulty of traditional detection methods to simultaneously achieve both high sensitivity and high directionality. To address this limitation, this work proposed a line-defect phononic crystal structure that enabled directional enhancement and detection of weak target signals under intense spatial noise interference by coupling defect-state localization characteristics with anisotropy mechanisms. Through theoretical derivation and finite element numerical simulation, the directional enhancement properties of this structure were systematically validated. Furthermore, numerical simulations were conducted to validate the detection of weak harmonic signals and weak bearing fault signals under strong spatial noise interference. The results demonstrated that this line-defect phononic crystal structure exhibited high feasibility and outstanding performance in detecting weak acoustic signals. This work provides novel insights for developing new acoustic detection methods combining high sensitivity with high directivity, showcasing unique advantages and broad application prospects in acoustic signal sensing, enhancement, and localization.

The present Special Issue on "Metamaterials and Their Devices, Second Edition" can be considered a status report reviewing the progress achieved over the past two decades in several subject areas affected by metamaterials and their devices.

https://doi.org/10.3390/cryst16020092

Conflicts of Interest: The author declares no conflict of interest.

References

1. Cheng, B.; Zou, Y.; Liang, T.; Ye, A.; Zhai, K.; Lv, L. A Long-Wave Infrared Circularly Polarized Photodetector Based on an Array of Trapezoidal Silicon Pillars. *Crystals* **2025**, *15*, 993. [CrossRef]
2. Deng, M.; Zou, Y.; Wang, H.; Wen, B. Wavelength Conversion in Photonic Crystal Fibers via Multiple Raman Redshifts and Soliton Spectral Tunneling. *Crystals* **2025**, *15*, 962. [CrossRef]
3. Zhang, S.; Mu, J.; Xiao, J.; Xu, H. Line-Defect Phononic Crystal Structure for Directional Enhancement Detection of Weak Acoustic Signals. *Crystals* **2025**, *15*, 907. [CrossRef]
4. Deng, J.; Wu, J.; Chen, X.; Zhang, X.; Li, S.; Song, Y.; Wu, J.; Xu, J.; Deng, S.; Wu, Y. Tandem Neural Network Based Design of Acoustic Metamaterials for Low-Frequency Vibration Reduction in Automobiles. *Crystals* **2025**, *15*, 676. [CrossRef]
5. Cheng, B.; Zou, Y.; Ge, Z.; Lv, L.; Liang, T.; Zhai, K.; Song, G. Design of Multifunctional Polarization Waveplates Based on Thermal Phase-Change Metasurfaces. *Crystals* **2025**, *15*, 462. [CrossRef]
6. Nasiri, M.; Wang, Y. Evolution of Phonon Spectral Energy Density in Superlattice Structures. *Crystals* **2025**, *15*, 446. [CrossRef]
7. Matei, A.T.; Visan, A.I.; Popescu-Pelin, G.F. Design and Processing of Metamaterials. *Crystals* **2025**, *15*, 374. [CrossRef]
8. Soroosh, M.; AL-Shammri, F.K.; Maleki, M.J.; Balaji, V.R.; Rajarajan, V.; Adibnia, E. A Compact and Fast Resonant Cavity-Based Encoder in Photonic Crystal Platform. *Crystals* **2025**, *15*, 24. [CrossRef]
9. Sun, X.; Yu, T.; Wang, L.; Lu, Y.; Chen, C. Embedded Rough-Neck Helmholtz Resonator Low-Frequency Acoustic Attenuator. *Crystals* **2025**, *15*, 12. [CrossRef]

https://doi.org/10.3390/cryst16020092

crystals

MDPI

Review

Design and Processing of Metamaterials

Andrei Teodor Matei [1], Anita Ioana Vișan [2,*] and Gianina Florentina Popescu-Pelin [2,*]

[1] IT Center for Science and Technology, 25 No. Av. Radu Beller Str., 011702 Bucharest, Romania; matei.andrei5@gmail.com

[2] National Institute for Lasers, Plasma and Radiation Physics, 077125 Măgurele, Romania

* Correspondence: anita.visan@inflpr.ro (A.I.V.); gianina.popescu@inflpr.ro (G.F.P.-P.)

Abstract: Metamaterials represent artificially structured materials that exhibit unusual properties, such as a negative refractive index, negative permeability and permittivity, negative cloaking by Poisson ratios and optical effects, etc., which are inaccessible in natural materials. According to recent developments, novel devices and tools based on metamaterials are attracting great interest as they offer improved performance, functionality, sensitivity, biocompatibility, complex structures, and design freedom. Leveraging numerical design approaches, such as finite element analysis and finite difference time domain methods, researchers have tailored metamaterials to meet specific requirements in various areas through a range of manufacturing techniques. These materials can be broadly classified into optical, mechanical, thermal, electromagnetic, and acoustic categories based on their properties and intended use. The choice of fabrication method depends heavily on the specific application, the desired scale, and the complexity of the metamaterial design. These manufacturing methods can be broadly divided into top-down and bottom-up approaches, while each of them has advantages and limitations and offers valuable pathways for the development of the final product. This review offers a basic overview of metamaterials, covering their fundamental principles, fabrication and characterization techniques, and current design methodologies. It also explores their diverse applications, including specific case studies in medicine, while addressing existing limitations and challenges. Finally, this review highlights future perspectives, emphasizing the need for continued innovation in fabrication and characterization to unlock the full potential of metamaterials.

Keywords: metamaterials; design methodologies; fabrication techniques; medical applications

1. Introduction

Metamaterials, whose properties rely more on structure than chemical composition, have emerged in the last two decades as a significant area of research due to their unique properties not typically associated with other classes of materials [1]. Metamaterials, through their carefully engineered subwavelength structure (i.e., structural features smaller than the wavelength of interacting waves), differ from conventional materials by exhibiting unconventional electromagnetic properties—such as negative refractive indices, negative electrical permittivity, and negative magnetic permeability [2]. While these properties are often achieved through artificial design, analogous phenomena are observed within biological systems. For example, the intricate nanostructures present in butterfly wings [3] and reptile skin [4] produce structural coloration and optical effects that resemble those of engineered metamaterials. In both cases, the observed properties stem not from the chemical composition alone but from the precise organization of substructures at subwavelength scales [5–7].

In the Greek language, the prefix "meta" translates to "beyond". Metamaterials represent a novel class of engineered materials that exhibit electromagnetic properties not found in natural substances. Typically, natural materials, such as diamond and glass, possess positive values for the refractive index, magnetic permeability, and electrical permittivity. In contrast, metamaterials can demonstrate negative refractive indices [8,9] negative electrical permittivities [10,11], and negative magnetic susceptibilities [12,13]. Metamaterials encompass a range of engineered materials, including those with a negative refractive index along with simultaneously negative permittivity and permeability, as well as materials with large negative thermal expansion or near-zero in-plane thermal expansion, large Poisson ratios, or reversed macroscopic magnetization [14]. These materials are often referred to by various terminologies, including left-handed (LH) materials [15,16] backward-wave (BW) media [17], negative-index materials (NIMs) [8], and double-negative (DNG) media [6,18].

Metamaterials exhibit unique features, such as perfect lensing [17], classical electromagnetically induced transparency [19–21] cloaking abilities [22], high-frequency magnetism [23] and dynamic modulation of terahertz (THz) radiation [24], and phenomena, like the reverse Doppler effect and the reverse Cerenkov effect [25]. These exceptional properties facilitate the creation of functional devices capable of switching and tuning [26–29]. Based on their permittivity and permeability, metamaterials can be classified as mu-negative materials (MNG), epsilon-negative materials (ENG), double-positive materials (DPS), and DNG materials [30]. MNG and ENG are considered single-negative materials, while double-negative and double-positive materials can be engineered for specific frequency bands.

The applications of metamaterials are far-reaching, impacting various fields, including optics, acoustics, and electromagnetics. In optics, they enable groundbreaking technologies, such as superlenses that exceed the diffraction limit [5,31]. In acoustics, metamaterials offer innovative solutions for sound insulation and noise control, highlighting their versatility [32]. Furthermore, the development of electromagnetics is a key driver in enhancing wireless communication technologies by improving antenna performance and enabling new functionalities in wireless devices [5,33].

Also, metamaterials hold transformative potential in medicine, particularly in enhancing wireless telemetry systems for early disease diagnosis and continuous physiological monitoring, as exemplified by their use in wireless endoscopes and advanced cancer detection devices that leverage changes in tissue permittivity and conductivity [34–36]. Furthermore, metamaterials improve medical imaging techniques, such as MRI, by significantly reducing scan times and enhancing signal-to-noise ratios [37], while their applications in microwave hyperthermia and wireless strain sensing contribute to more effective cancer treatments and patient monitoring [38,39].

The study of metamaterials is not a recent development; rather, it has a rich historical foundation. The fundamental idea of developing materials with tailored properties through microstructural design dates back nearly a century [5,31].

The concept of metamaterials, artificial materials engineered to exhibit properties not found in nature, has its roots in the pioneering work of Russian physicist Victor G. Veselago in the 1960s. Veselago first theorized the existence of materials with a negative refractive index, proposing that such materials could exhibit unique electromagnetic properties, including reversed-phase velocity and energy flow [15]. His groundbreaking work laid the theoretical foundation for metamaterials, although the practical realization of these materials remained elusive for decades [40].

Veselago's theoretical predictions were based on Maxwell's equations, which describe the behavior of electromagnetic waves. He demonstrated that in a hypothetical material with simultaneously negative permittivity and permeability, the phase velocity of light would be antiparallel to the direction of energy flow, a phenomenon not observed in natural materials. Despite the elegance of his theory, Veselago's ideas were initially met with skepticism, as no known materials exhibited the required negative refractive index at the time [41].

The field of metamaterials gained traction in the late 20th century, thanks to advances in materials science and nanotechnology. In 1968, Leonard introduced parallel plate wave transmission structures, which provided a practical framework for exploring wave behavior in engineered materials [42]. That work marked an early step toward the experimental realization of metamaterials, as it demonstrated the potential for manipulating electromagnetic waves using structured materials.

The true breakthrough in metamaterials research came in the early 2000s when scientists successfully fabricated materials with negative refractive indices. This achievement was made possible by the development of split-ring resonators (SRRs) and other nanostructured components, which allowed for precise control over electromagnetic properties [30]. These advances were documented in several key studies, including a comprehensive review by Pendry and Smith [43], which highlighted the potential of metamaterials for applications such as superlenses, cloaking devices, and advanced antennas.

Since then, the field of metamaterials has expanded rapidly, with research focusing on both fundamental principles and practical applications. A review by Engheta and Ziolkowski [28] explored the theoretical and experimental progress in metamaterials, emphasizing their potential to revolutionize photonics, telecommunications, and sensing technologies. The history of metamaterials (Figure 1) is a testament to the power of theoretical insight and technological innovation. From Veselago's initial proposal to the cutting-edge applications of today, metamaterials have evolved from a theoretical curiosity to a cornerstone of modern materials science.

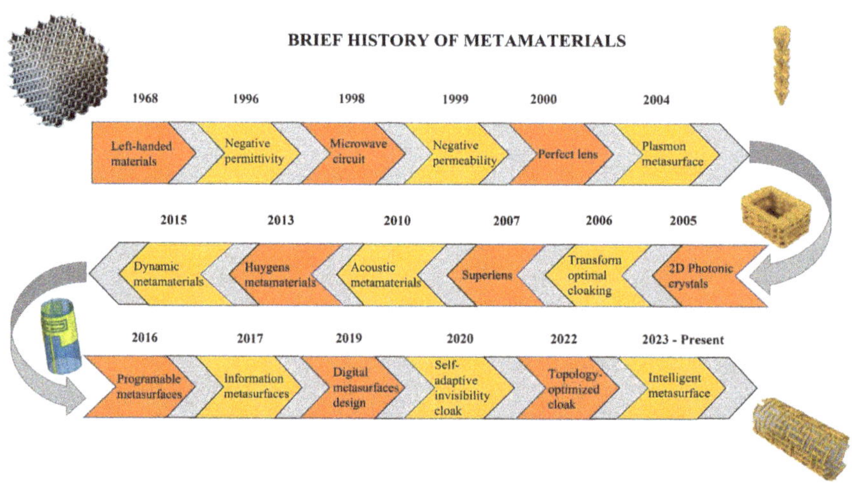

Figure 1. Brief history of metamaterials (readapted from [44–46]).

As research continues, the potential for metamaterials to enable new technologies and deepen our understanding of wave–matter interactions remains vast.

While initial cloaks were narrowband, recent work on dielectric metamaterials [47] and quasi-conformal mappings [48] addresses losses and scalability, paving the way for clinical applications.

This review summarizes the state-of-the-art design and processing of metal, ceramic, and semiconductor metamaterials while outlining promising avenues for future fruitful research.

1.1. Fundamental Principles of Metamaterials

A thorough understanding of natural materials enhances the design of metamaterials, enabling the development of new, controllable materials [49]. Metamaterials are known for the specially designed unit cells or atoms (periodic or nonperiodic subwavelength "meta-atoms" [49]) that work together to exhibit unconventional properties not found in natural materials, making their field inherently interdisciplinary by bridging areas such as electromagnetics, optics, solid-state physics, and acoustics [50,51]. Thus, three distinct categories of metamaterials, i.e., homogeneous (periodic), random, and inhomogeneous, have been developed at microwave frequencies, each exhibiting unique structural arrangements of meta-atoms and facilitating the exploration of novel physical phenomena and the advancement of practical devices [49,52].

Due to the availability of extreme medium parameters, early studies of homogeneous metamaterials revealed their potential to realize unusual physical phenomena such as negative permittivity, negative permeability, and zero index of refraction [49,53].

A highly homogeneous zero-index metamaterial (ZIM) in the microwave regime based on a $SrTiO_3$ ceramic was proposed, engineered, and characterized by Liu et al. Their study introduced and experimentally validated a novel antenna design that incorporated the developed ZIM within a metallic waveguide, achieving enhanced directivity. The fabricated prototype achieved a peak directivity of 11.2 dB, despite its compact aperture dimensions of $1.2\lambda_0 \times 1.2\lambda_0$. Furthermore, the antenna demonstrated sustained high directivity across a broad range of aperture sizes, from $0.5\lambda_0 \times 0.5\lambda_0$ to significantly larger scales, effectively approaching the theoretical directivity limit over this extended range [54].

On the other hand, NIMs are applied in superlenses to achieve resolutions below the diffraction limit, making them invaluable for medical diagnostics [55].

The integration of NIMs in imaging systems has shown promising results in visualizing nanoscale structures, which traditional lenses cannot achieve [56]. Wang Y. et al. proposed the integration of a NIM lens with a microstrip antenna, enhancing its directional radiation capability, relative bandwidth, and average gain by 7.54 dB, with a maximum gain of 10.24 dB in the 18.38–20.66 GHz frequency range [56].

Random metamaterials, with randomly packed meta-atoms, are composed of conductive and insulating phases, where negative permittivity is achieved through an interconnected conductive network, allowing for enhanced control over electromagnetic properties, including radiation, reflection, and scattering. The generation of electromagnetic diffusion or Lambertian reflection is a key application of this approach, where varying the geometrical parameters of meta-atoms across space leads to spatially random distributions of effective medium parameters such as permittivity, permeability, the refractive index, and wave impedance [49,57]. The ability to fabricate custom-made magneto-dielectric materials has led to the development of transformation optics, inspired by Einstein's theory of general relativity.

Gholipur and Bahari demonstrated that silver nanowires, randomly dispersed within a $Zr_{0.9}Ni_{0.1}O_Y$ dielectric matrix, exhibited both negative permittivity and negative magnetic permeability within the microwave spectral range. The selection of silver as the metallic component was based on its minimal loss factor, a crucial consideration for metamaterials operating in the microwave to infrared frequency range. Furthermore, the use of a dielectric host with a high permittivity was deemed advantageous for this composite material [58].

Another study was conducted on iron/epoxy random metamaterials, synthesized through a simple mixing and pressure-forming technique, to investigate the influence of iron content on their electrical properties. The composites demonstrated a percolation threshold at an iron volume fraction of 0.425. Below this threshold, conductivity was governed by hopping conduction due to the isolated nature of the iron particles, while above, a conductive network formed, leading to metal-like conduction and negative permittivity [59].

This approach maps theoretical spatial distortions onto real-world inhomogeneous and anisotropic metamaterials, offering precise control over electromagnetic wave propagation. Inhomogeneous metamaterials, characterized by nonperiodic meta-atom distributions, provide enhanced flexibility in manipulating electromagnetic fields.

The design of inhomogeneous electromagnetic metamaterials (IEMMMs) for wireless power transfer systems shows how varying effective permeability can enhance efficiency by up to 28%. IEMMMs are tailored for specific applications, such as improving magnetic field modulation in power transfer systems. In An et al.'s paper, an inhomogeneous EMMM (IEMM) with three kinds of effective permeability was proposed to improve the range and efficiency of a wireless power transfer system (WPTS). Their study examined the effective permeability of circular, square, and "8"-shaped coil unit cells and investigated the magnetic field control achieved by negative permeability mediums. Based on the analysis, a two-dimensional Impedance-Enhanced Magnetic Metamaterial (IEMMM), composed of square and "8"-shaped coil unit cell combinations, was developed. Finite element simulations confirmed that the IEMMM effectively modulated the magnetic field and reduced flux leakage in a Wireless Power Transfer System (WPTS). Notably, the designed IEMMM enhanced the power transmission efficiency of the WPTS by as much as 28% [60].

This design freedom, combined with subwavelength geometries and gradient variations, allows for the creation of novel devices, such as invisibility cloaks and advanced optical lenses, driving innovation in electromagnetic wave control and engineering applications [49,59,61].

The world of electromagnetic waves, including light, displays many peculiar effects compared with Newton's mechanics, particularly due to their wave-like nature, finite wave speeds, and dependence on parameters like electric permittivity and magnetic permeability [62,63].

1.2. Electromagnetic Theory

Electromagnetic phenomena are primarily governed by Maxwell's equations that describe the relationships between electric and magnetic fields, their sources, and the properties of materials [64]. The electromagnetic properties of metamaterials are intricately linked to their ability to alter the propagation of electromagnetic waves through engineered structures [64]. This manipulation is often described in terms of permittivity (ε) and permeability (μ), which characterize a material's response to electric and magnetic fields, respectively. The refractive index ($n = \sqrt{\mu\varepsilon}$) and impedance ($Z = \sqrt{\mu/\varepsilon}$), considered macroscopic effective parameters, are derived quantities, directly dependent on permittivity and permeability. Consequently, alterations in either ε or μ will induce corresponding changes in n and Z in a coupled manner, as these relationships are fundamentally linked through Maxwell's equations [30,64]. In metamaterials, these parameters can be designed to achieve negative values, leading to a negative refractive index. The capacity to control this parameter leads to novel wave interactions, such as reversed Doppler effects and negative refraction, which have significant implications for optical imaging and cloaking technologies [65,66].

Furthermore, it is crucial to recognize that non-classical permittivity and permeability are usually frequency-dependent, often occurring within narrow spectral ranges [67]. These properties are particularly relevant in regions such as visible light or nearby frequencies, where the manipulation of electromagnetic waves is most impactful [68].

The theoretical framework for understanding metamaterials is often based on transformation optics, which provides a mathematical approach to designing materials that can control the flow of electromagnetic waves in desired ways. Thus, from a coordinate transformation, researchers can create a perfect invisibility cloak, similar to a "mirror" that projects the originally flat electromagnetic space into physical space [69–71].

Metamaterials can be classified using various schemes based on their electromagnetic properties and applications [72].

The practical impact of transformation optics was first demonstrated in 2006 with a microwave invisibility cloak. By spatially varying ε and μ using an SRR, the cloak bent microwaves around a hidden object, validating Veselago's theoretical predictions and Pendry's superlens concept [73].

Figure 2 represents the primary classification of metamaterials based on their permittivity and permeability [1]. Integrating topology optimization into the design can lead to a better configuration of metamaterials that fits in the area of permittivity and permeability. The unit cell geometry (e.g., split-ring resonators or wire arrays) determines these effective medium properties, allowing tailored wave manipulation for applications such as cloaking and antennas.

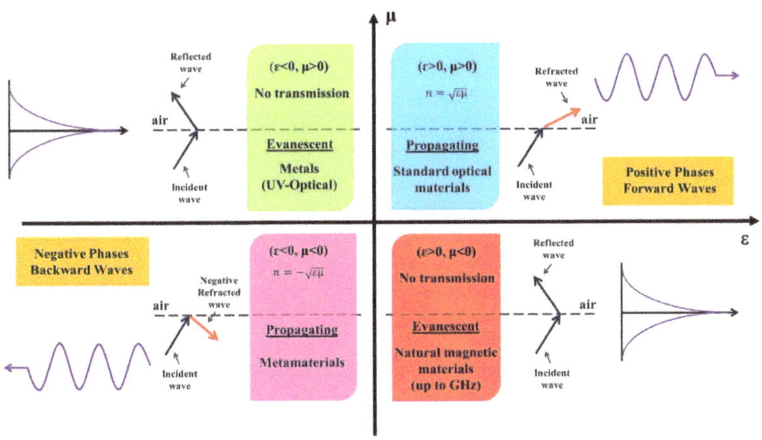

Figure 2. Classification of metamaterials based on their permittivity (ε) and permeability (μ) (readapted from [1,72]).

The figure illustrates a quadrant-based graphical representation, where each quadrant delineates a specific combination of ε and μ values, highlighting the unique electromagnetic properties of each class, as follows:

- Double-Positive (DPS): both ε and μ are positive (top-right quadrant), typical of conventional materials like dielectrics.
- Epsilon-Negative (ENG): $\varepsilon < 0$, $\mu > 0$ (top-left quadrant); found in plasmonic materials (e.g., metals at optical frequencies).
- Mu-Negative (MNG): $\varepsilon > 0$, $\mu < 0$ (bottom-right quadrant); exhibited by certain magnetic resonators.
- Double-Negative (DNG): both ε and μ are negative (bottom-left quadrant), enabling negative refraction and superlensing.

Recent advancements in the field have also highlighted the role of plasmonic effects in enhancing the performance of metamaterials. Plasmonic metamaterials leverage the interaction between light and free electrons in metals to achieve strong field localization and enhanced nonlinear optical responses [74]. This has opened new avenues for applications in sensing, imaging, and information processing, where the ability to control light at the nanoscale is crucial [75]. Moreover, the integration of metamaterials with microelectromechanical systems (MEMSs) has emerged as a promising strategy for creating reconfigurable devices. This integration allows for dynamic control over metamaterial properties, enabling applications that require adaptability to changing conditions [75].

The development of plasmonic biosensors for early disease detection represents a significant advancement in nanotechnology, particularly through the application of plasmonic metamaterials. These biosensors utilize gold or silver nanostructures to support localized surface plasmon resonances (LSPRs), thereby enhancing both sensitivity and specificity in disease biomarkers detection. The interaction of light with these nanostructures generates intense electromagnetic fields that enable the detection of biomolecules at extremely low concentrations [76] making them suitable for applications in cancer diagnostics (detecting HER2 for breast cancer) and infectious disease screening (identifying pathogens like SARS-CoV-2) [77] or point-of-care testing when their miniaturized design allows for use in resource-limited settings [78].

The excitation of surface plasmons at the metal–dielectric interface leads to a measurable shift in resonance wavelength or intensity when target biomolecules are present [76].

These sensors can detect biomolecules down to picomolar levels, enabling the real-time monitoring of interactions [79].

While plasmonic biosensors show great promise, challenges such as reproducibility and scalability remain, which could hinder their widespread clinical adoption.

As the field continues to evolve, the interplay between electromagnetic theory and practical applications of metamaterials will undoubtedly lead to further innovations and discoveries.

1.3. Justification of Review Importance

This review is primarily intended for researchers, engineers, and academics within the fields of materials science, physics, and engineering. A comprehensive overview of this rapidly evolving field is crucial because it synthesizes the extensive and rapidly evolving body of knowledge surrounding metamaterials, which possess unique properties that can be leveraged for groundbreaking applications in medicine, optics, telecommunications, and energy harvesting. By critically analyzing the current literature in the field, this review will shed light on prevailing trends, pinpoint existing gaps, and highlight challenges that need to be addressed in the field. Such insights will offer valuable guidance for future research initiatives and foster meaningful collaborations. Moreover, this work will serve as a valuable resource for professors and students, enhance academic discourse, and stimulate technological advancements across various industries.

Recent advancements in artificial intelligence (AI) and additive manufacturing (AM) have significantly impacted the design and fabrication of metamaterials. AI-driven techniques, particularly machine learning (ML) and deep learning (DL), facilitate the inverse design of metamaterials by predicting optimal geometries and configurations tailored to specific performance criteria. This overview discusses significant methodologies and their implications in the field. This review explores how AI techniques, including Generative Adversarial Networks (GANs), Variational Autoencoders (VAEs), and deep learning, work alongside additive manufacturing to streamline and enhance the design and prototyping of metamaterials. While the integration of AI and AM in metamaterial design presents numerous advantages, challenges remain, such as the need for more interpretable models and the handling of complex loading scenarios in practical applications. Overcoming these issues is essential for the ongoing development of metamaterials. Moreover, this review also highlights the critical advancements and challenges in utilizing additive manufacturing techniques, including fused deposition modeling (FDM), stereolithography (SLA), and selective laser sintering (SLS), in producing intricate metamaterials, particularly for medical applications. It underscores the importance of addressing issues related to scalability and quality consistency to ensure the broader adoption of these technologies. These innovations pave the way for metamaterials to transition from experimental concepts to commercially viable solutions. By encompassing these advancements, this review not only provides a snapshot of the current state of metamaterials but also highlights the transformative potential of AI and AM in pushing the boundaries of material performance. Future research directions include the development of self-learning metamaterial systems that autonomously adapt to operational conditions, as well as the exploration of hybrid fabrication techniques that integrate both top-down and bottom-up approaches for enhanced functionality. As the field continues to evolve, this review serves as a fundamental resource for guiding interdisciplinary efforts and unlocking the full potential of metamaterials in real-world applications.

2. Principles of Metamaterial Design

The design of metamaterials serves as a crucial bridge between theoretical concepts and practical applications. Functionally, metamaterials include electromagnetic metamaterials (manipulating electromagnetic waves, e.g., cloaking devices [72], acoustic metamaterials (controlling sound waves, e.g., acoustic cloaking) [80], mechanical metamaterials (exhibiting unusual mechanical properties like negative Poisson's ratios or tunable stiffness) [81,82] thermal metamaterials (managing heat flow, e.g., thermal cloaking) [83], and optical metamaterials (manipulating light, e.g., superlenses) [8]. Based on scale, they range from macroscopic (visible structures for engineering applications) to microscopic/nanoscopic (used in photonics or nanomedicine [84]. Design-wise, they can be periodic (repeating unit cells) or aperiodic/random (irregular structures for specialized uses) [85]. Mechanical metamaterials, in particular, derive their unique properties—such as auxetic behavior (expanding when stretched), negative compressibility, or programmable deformation—from their carefully designed internal architectures rather than their chemical composition. Their design relies on unit cell geometry, multiunit cell arrangements, and material selection, enabling applications in lightweight structures, energy absorption, and advanced robotics [84].

Traditionally, the design process involves extensive full-wave electromagnetic simulations and the retrieval of scattering (S) parameters, in which numerical simulations are iteratively conducted to fine-tune the constitutive parameters of individual metamaterial elements [50]. The design of metamaterials is grounded in several fundamental concepts, including NIMs, anisotropic materials, and resonant structures. Negative-index materials lead to a modified Snell's law (the refracted angle becomes "negative", altering the traditional calculation of the light's refraction when it passes between two media with different refractive indices) [28,64], enabling applications such as cloaking, superlensing, and sensors [5,86]. These NIMs are typically regarded as artificially engineered materials exhibiting unique optical properties as a result of combining two or more elemental materials [87]. Anisotropic materials exhibit direction-dependent properties (manipulation of wave propagation), which can be tailored for specific applications in optics [88] and acoustics [89]. Furthermore, anisotropic acoustic metamaterials can convert evanescent waves onto propagating waves (by effectively coupling large wave vector components), allowing high-resolution imaging even in the far field [5,31,80]. Subwavelength resonant structures are the key components of metamaterials (can respond, as a homogenous material, to the electric and magnetic field), giving them unique properties [49,90]. A typical highly conductive metamaterial structure is the SRR, which exhibits a balance between its capacitance and inductance and can be engineered to be very compact, having dual band features [91,92]. Based on the resonant properties of the structure and when applying a time-varying magnetic field perpendicular to the surface (generates currents), an SRR creates a magnetic field that can either counteract or reinforce the original field, leading to either positive or negative permeability (μ) [5,86,93]. Resonators can be classified as internal and external depending on their position towards the structure. Internal resonators are an integral part of the material and, as a matter of fact, do not require additional space. On the other hand, external resonators are placed externally, which protects the main structure and is an accessible way to be repaired [94]. Metamaterials can provide various types of resonances depending on other compositions and interactions with other waves: band-gap resonance, acoustic resonance [94], plasmonic resonance [95], and Fano resonance [96]. Dispersion is a fundamental characteristic of electromagnetic systems originating from the interaction between light and materials. Metamaterials exhibit design flexibility due to their artificial structures, enabling precise manipulation of electromagnetic properties at subwavelength scales [97]. This unique characteristic enables the intrinsic resonance of subwavelength structural elements to generate significant dispersion in their effective material parameters. Metamaterials can overcome the inherent limitations of classic diffraction, reflection, and refraction theory, paving the way for advanced dispersion engineering in a wide range of applications [98]. Design methodologies for metamaterials encompass analytical approaches, numerical methods, such as the finite-difference time-domain (FDTD) and the finite element method (FEM), and experimental validation techniques [71,99]. Numerical simulations are particularly valuable for modeling complex interactions within metamaterials, allowing for optimization prior to physical fabrication [71]. The FDTD method, a simple and versatile technique, has been widely used to simulate electromagnetic wave propagation (by directly implementing Maxwell's time-dependent curl equations [100] in metamaterials by treating them as dispersive media) [78,95,99,101].

An alternative numerical method for solving the time-domain Maxwell's equations in the presence of metamaterials is the FEM, which is used for modeling complex geometries [102]. This approach is suitable for settling heat conduction problems, especially for thermal metamaterials, and allows us to understand the damages caused in the matrix of a composite material that is exposed to extreme conditions, including different temperatures, pressures, and static or dynamic loading [103,104]. While FDTD and the FEM are the main methods for electromagnetic analysis, other numerical techniques, including Rigorous Coupled Waveguide Analysis [105], Method of Moments [106,107], and the Boundary Element Method [108], are also employed in certain situations. Design optimization strategies for metamaterials should be flexible enough to integrate with any of these numerical modeling methods. Different experimental validation techniques are used to provide theoretical or calculated predictions about the behavior and properties of metamaterials. These techniques can be classified depending on the type of response that they are designed to measure, e.g., electromagnetic [109], mechanical [110], or acoustic [111], and are essential for validating metamaterial designs.

2.1. Metamaterials Design and Fabrication Techniques

Metamaterials research has expanded significantly in the last decade, spanning from the basic pursuit of designing new materials to the investigation of the unique properties emerging from their subwavelength structure. A range of capabilities is required to design and analyze these complex material archetypes. This section presents a brief overview of several design and fabrication methods currently in use or under development. The subwavelength nature of single metamaterial spacings offers the possibility of tailored responses, with the potential for significant effects on properties in electromagnetic and other functions. Hybrid structures that allow for foreshadowing of system properties are also considered.

Figure 3 illustrates the design space of mechanical metamaterials, emphasizing the key parameters that influence their design and performance. These parameters are derived from three main categories: (1) materials and fabrication techniques, which determine the physical and structural properties of the metamaterials; (2) unit cell design, which defines the individual behavior of each repeating structural element; and (3) multiunit cell architecture, which governs the overall arrangement and interaction of multiple unit cells to achieve desired mechanical properties at a larger scale. This comprehensive framework provides a structured methodology for designing mechanical metamaterials, allowing researchers to tailor their properties for specific applications, such as lightweight structures, energy absorption, or tunable stiffness. By understanding and optimizing these parameters, it becomes possible to push the boundaries of material performance and explore new functionalities in engineering and technology.

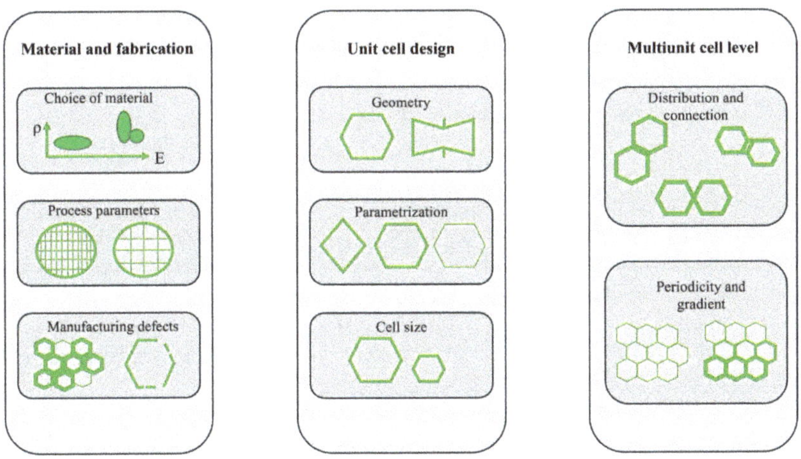

Figure 3. Design space of mechanical metamaterials: parameters originating from material and fabrication, unit cell design, and multiunit cell level architecture (readapted from [112]).

A major class of well-known properties is the permittivity and permeability of electrical materials, often expressed in tensor form. From these properties, Maxwell's equations are expressed in a form with material properties. A second class is the scattering parameters, which relate either to the tensor properties or to transfer matrices that represent the boundaries of the material. Additional tool sets that enable the optimization of specific response properties provide methods that could also exploit constitutive requirements, such as those derived from transformative electromagnetic techniques. Additionally, targeted manufacturing options, such as lithography, etching, and screen printing, can be employed. These techniques can be broadly categorized into computational methods, fabrication approaches, and hybrid systems that integrate multiple functionalities.

Top-down methods like lithography enabled early transformation optics devices, such as the Duke microwave cloak [48]. However, bottom-up approaches (e.g., 3D printing) [113] now allow more complex gradient-index designs for broadband applications.

The advent of computational methods has significantly transformed the design landscape of metamaterials. Techniques such as topology optimization, evolutionary algorithms, and deep learning have emerged as powerful tools for the inverse design of metamaterials. For instance, deep generative models, like Generative Adversarial Networks (GANs) and Variational Autoencoders (VAEs), streamline the design process by predicting and generating metamaterial structures that meet specific performance criteria, thus reducing the reliance on trial-and-error approaches [114]. These methods allow for the exploration of complex design spaces, enabling the discovery of novel configurations that exhibit desired electromagnetic responses.

Employed to streamline the design process, GANs reduce reliance on trial-and-error methods, enabling rapid exploration of design spaces [115].

These models assist in generating new metamaterial designs by learning from existing data, thus enhancing the efficiency of the design process [116].

Techniques such as image-based deep learning networks minimize preprocessing steps, allowing for direct generation of metamaterial designs from input parameters [117].

On the other hand, additive manufacturing techniques, including 3D printing, have revolutionized the fabrication of metamaterials by allowing for the creation of intricate geometries that were previously unattainable through traditional methods. This approach enables the production of metamaterials with tailored properties, such as customizable thermal expansion and mechanical responses [118]. Furthermore, the integration of stimuli-responsive materials with 4D printing techniques facilitates the development of metamaterials that can adapt their properties in real time based on environmental changes [118].

The flexibility and precision of additive manufacturing also support the fabrication of complex multimaterial structures, enhancing the functional capabilities of metamaterials [119].

The exploration of hybrid systems that combine different types of metamaterials or integrate active components into passive structures is gaining traction. For example, the incorporation of digital logic gates into soft, conductive mechanical metamaterials allows for the creation of multifunctional devices that can perform computational tasks while also exhibiting mechanical properties [120]. This modular approach not only enhances the versatility of metamaterials but also paves the way for the development of intelligent systems capable of self-powered information processing [121].

Advanced simulation techniques, such as discrete dipole approximation and finite element analysis, play a crucial role in predicting the behavior of metamaterials under various conditions. These methods enable researchers to model complex interactions within metamaterials and optimize their designs before fabrication [122]. Additionally, the use of machine learning algorithms to analyze simulation data can further refine design parameters and improve the accuracy of predictions regarding metamaterial performance [123].

As the field of metamaterials continues to evolve, new design techniques are emerging that leverage advancements in materials science and engineering. For instance, the development of programmable granular metamaterials allows for the tuning of mechanical properties through structural design, offering a vast design space for energy absorption applications [124]. Furthermore, the integration of artificial intelligence in the design process is expected to enhance computational efficiency and accelerate the discovery of novel metamaterials with tailored functionalities [123]. The design techniques for metamaterials are diverse and continually advancing, driven by the need for innovative solutions across various applications. The combination of computational methods, advanced fabrication techniques, and hybrid systems is paving the way for the next generation of metamaterials with unprecedented capabilities. The most comprehensive coverage of the design and fabrication of metamaterials would include elements from all of these classes.

2.2. Unit Cell Design

The unit cell in metamaterials is fundamental as it encapsulates essential information about the crystal structure, including geometric attributes, material characteristics, and lattice spacing. These elements are pivotal in establishing the periodicity of the crystal and influencing the position and extent of the band gap [125]. The repeating unit cell establishes the crystal periodicity and controls the position and extent of the band gap [125]. Choosing the appropriate unit cell shape is fundamental in metamaterial design, as it directly influences the resonant bandwidth, incident-angle stability, and cross-polarization level. Given their structures, these shapes can be broadly classified as centrally connected N-poles, rings, internal solids, and composites [126]. The foundational unit cell supports the design of a reconfigurable metamaterial device and determines the fundamental behavior of the structure. Typically, the design process begins with a static configuration, which is then enhanced by incorporating tunable elements, materials, or structures [127]. The unit cell's physical geometry controls the electromagnetic coupling within the metamaterial, introducing frequency selectivity with either desired or undesired effects, and can facilitate the suppression of either electric or magnetic excitations. Moreover, the unit cell geometry can be engineered to couple electric and magnetic responses strongly [127].

The design of metamaterial unit cells is a multifaceted challenge that includes considerations of geometry, material selection, and operational context. For example, any changes to the physical dimensions of an SRR that affect either inductance (L) or capacitance (C) will directly impact its resonant frequency [128]. Thus, a narrower gap increases capacitance, which lowers the resonant frequency. Conversely, a wider gap decreases capacitance, raising the resonant frequency. The overall size and shape of the rings influence the inductance. Larger rings generally result in higher inductance. Using nested or multiple rings introduces more complex interactions, affecting both L and C and allowing for the creation of multiband resonances [129].

Asymmetric designs, such as L-shaped or zigzag wires, introduce anisotropy, enabling polarization-dependent responses. In contrast, stacked nanowire unit cells in hyperbolic metamaterials create extreme anisotropy, where the permittivity tensor has opposite signs for different components, enabling subdiffractional imaging. Asymmetric unit cell designs disrupt the symmetry of the metamaterial, resulting in its electromagnetic response varying depending on the polarization of the incident light. Specifically, symmetric unit cell designs disrupt the symmetry of the metamaterial, resulting in its electromagnetic response varying depending on the polarization of the incident light [130]. This is crucial for applications requiring precise control over polarization, such as optical isolators, polarization converters, and chiral metamaterials. Asymmetric electromagnetic wave transmission of linear polarization is possible via polarization conversion through chiral metamaterial structures [131].

Coupling electric LC resonators with magnetic SRRs enables the creation of double-negative media. All-dielectric unit cells, such as spheres or disks of high-index dielectrics, support Mie resonances with spectral positions that depend on the particle diameter and aspect ratio [132]. Reconfigurable unit cells, like MEMS-actuated SRRs or liquid crystal-filled gaps, enable dynamic switching of EM properties, though complex geometries often require advanced lithography and present fabrication trade-offs [28].

Effective manipulation of the unit cell can lead to significant advances in the performance and applicability of metamaterials in various fields, including medicine [133], telecommunications [134], and even sensing technologies [135]. The unit cell is the basic "building block" of metamaterials [119], and its design can significantly influence the overall electromagnetic properties of the material. Therefore, it is necessary to first define what a unit cell is in the context of metamaterials. Although in many practical implementations of metamaterials, this takes the form of a volume of material with certain exceptional properties within a dielectric, the term "building block" is generally not useful or instructive for the design process of metamaterials [119]. In traditional crystalline materials, translational invariance to symmetry is the fundamental requirement of the unit cell. In other words, there are an infinite number of unit cells that are translated away from each other, so that the crystal is invariant when translated through each of these units. Theoretically, the same requirement applies to three-dimensional periodic materials as to metamaterials. However, for a practical realization of a metamaterial, the unit cell size is always limited by the size of the system to be realized [136].

In this sense, the unit cell is an imposed limitation, and its design generally lacks complete translational invariance to symmetry. The attainment of perfect translational invariance in practical metamaterials is significantly impeded by several key challenges, including spatial disorder arising from fabrication imperfections like nanoscale misalignments [130] and finite size effects, which cause edge effects [78,133]. Finite metamaterial samples exhibit mode truncation effects that are absent in infinite-lattice theories, while dispersive and lossy resonant elements challenge homogenization due to spatial dispersion and frequency-dependent losses [137], near-field coupling between unit cells at subwavelength scales leads to unintended interactions and "hotspots", and scalability versus performance trade-offs force compromises between large-area fabrication precision and throughput [8]. Researchers mitigate these issues through statistical homogenization to model disorder, edge optimization to minimize reflections, and active tuning to correct inhomogeneities using technologies like MEMS or liquid crystals dynamically [138,139].

The design of unit cells in practical metamaterials is, therefore, different from that in traditional materials, and it is often difficult to determine the symmetry of the material from a simple examination of the underlying configuration of basic elements [136]. This viewpoint is useful across all scales of metamaterial design, from simple wires on a printed circuit board that defines frequency-selective surface to subwavelength skin depth thickness penetration limits within the unit cells of acoustical metamaterials [140]. However, most designs are formed from an arrangement of a small number of basic building blocks, with the liquid crystal (LC) used in the design frequency for dimensional simplicity. The most common building blocks used to form electrical/magnetic and chiral topologies within metamaterials are the split-ring resonator and the square split-ring resonator, the Jerusalem or cross resonator, thin wires for electrically shorted reactions, split Cube geometries, and fishnets metamaterials and arrays of these topologies that impart an anisotropic response to the medium [127].

Khansanami et al. showed that the shape and thickness of the unit cell are critical for defining a metamaterial's mechanical and vibrational properties [136]. In another example, a square-shaped unit cell with specific dimensions was analyzed to optimize its metamaterial properties [141].

One of the primary considerations in unit cell design is the method of homogenization, which is essential for predicting the effective properties of metamaterials. The predictive accuracy of conventional homogenization methods, such as volumetric averaging, encounters significant limitations when predicting metamaterial effective properties. These include mismatches when the unit cell size approaches operating wavelengths [137], the failure to account for localized field discontinuities and enhancements at inclusion boundaries [49], inadequate representation of strong anisotropy and frequency dispersion [63], inability to model dynamic changes in nonlinear or tunable metamaterials [16], and neglect of near-field interactions and coupling between unit cells [30], especially in complex subwavelength structures or those with thin metallic inclusions [44,63,137,142]. Such considerations are crucial for ensuring accurate predictions of the metamaterial's optical properties. The choice of unit cell geometry significantly determines the operational bandwidth and resonant frequencies of metamaterials. For instance, an SRR unit cell design can significantly alter the current path, leading to new resonant frequencies, particularly favorable for multiband antenna applications [143]. Similarly, nested U-ring resonators have been shown to facilitate miniaturization while achieving multiband behavior, demonstrating that more complex geometries can enhance performance beyond traditional designs [144]. In contrast to metallic metamaterial unit cells, which are often complex, particularly when a magnetic response is essential, all-dielectric metamaterials (ADMs) can be employed with much simpler geometries (spheres and cubes) [127]. These findings underscore the necessity of innovative geometrical configurations in unit cell design to achieve desired electromagnetic properties. Thus, the integration of materials, such as LCs, into metamaterial designs has opened new opportunities for tuning and dynamic control of electromagnetic responses. Savo et al. presented an LC metamaterial absorber that can be electronically controlled to shape the spatial distribution of terahertz radiation, highlighting the potential of combining different materials within the unit cell to improve functionality [145]. This approach aligns with the trend towards reconfigurable metamaterials, where MEMSs are employed for independent tuning of multiple resonances at terahertz [146,147].

Integration of LCs into metamaterials enables dynamic control and adaptive properties [145] through real-time reconfigurability, enhanced tunability, and multifunctionality. Specifically, LCs' anisotropic refractive indices facilitate electrically tunable optical and THz metamaterials with dynamic switching capabilities [148]. Their dielectric property variations allow for the development of reconfigurable microwave [149] and radio frequency (RF) devices for beam-steering and frequency-agile sensing. Furthermore, LCs amplify nonlinear effects for ultrafast optical switches and dynamic light focusing, and their thermotropic phase changes allow for thermal and mechanical adaptability in applications like self-regulating thermal cloaks and soft robotics [150]. While response time limitations exist, innovations in hybrid LC–polymer composites and future integration with AI-driven design and nanoscale LC patterning hold promise for enhancing metamaterial functionality, thereby transforming static metamaterials into adaptive systems for diverse applications [146].

Such advancements highlight the importance of the geometric design of unit cells and the materials used in their construction.

2.3. *Topology Optimization*

Topology optimization (TO) is an effective and versatile design method developed decades ago, as a mathematical approach for optimizing structural shapes and material distribution in components, to maximize performance according to relevant design specifications [151–154]. Originally developed for structural mechanics problems [155], TO offers a systematic and performance-driven framework for designing metamaterials by iteratively optimizing the structural layout to achieve a specific objective (e.g., maximum stiffness, thermal dissipation, or minimum weight/cost) under defined constraints [156]. In the design of metamaterials, optimization is used to identify the optimal topology of the periodic unit cell microstructure [157]. Topology optimization, applicable to diverse physical phenomena including heat transfer and wave propagation (e.g., sound and electromagnetic wave), is now being employed to design metamaterials with a wide range of tailored properties, such as elasticity, viscoelasticity, extreme thermal expansion, photonic and phononic bandgaps, thermal exchange, and even negative permeability [157,158].

In the context of microstructure topology optimization, determining the effective properties of unit cells typically requires the application of numerical homogenization, which involves solving linear equilibrium equations. This process, however, is computationally intensive. To mitigate this, reduced-order modeling techniques that utilize Krylov subspaces have been proposed, enhancing computational efficiency. These methods employ an iterative solver to address large linear systems, leveraging a search space based on previous solutions. The approach assumes that the numerical stiffness matrices between successive iterations change only incrementally, reflecting small modifications in the structure's topology or shape [159].

Topology optimization methods like SIMP and SKB significantly influence metamaterial unit cell design [152]. SIMP, through penalizing intermediate densities, produces clear, high-contrast geometries ideal for static, high-stiffness applications like mechanical metamaterials, with computational efficiency but potential numerical artifacts [160], while SKB, using gradual stiffness reduction, creates smoother, more manufacturable structures suitable for complex, multimaterial designs like thermal or acoustic metamaterials, albeit with higher computational costs [161]. Hybrid approaches combining SIMP and SKB [162] leverage both clarity and smoothness, and the choice between them depends on the target application, with SIMP favored for precision [163] and SKB for adaptability in multi-physics designs. This impacts performance, manufacturability, and functionality uniquely [163] and provides ongoing advances in computational tools, expanding optimization possibilities [155].

Among the various methods available, the Solid Isotropic Material with Penalization (SIMP) method and the Soft-Kill Binary (SKB) method have gained significant attention due to their unique capabilities in generating optimal topologies [154,164]. The SIMP method excels in generating clear and distinct topologies, making it a popular choice for various engineering applications. On the other hand, the Soft-Kill Binary method offers smoother transitions and more robust optimization results, particularly in complex and multimaterial structures [154,164].

The SIMP method is one of the most widely used methods in topology optimization. It works by introducing a material model that penalizes intermediate densities, encouraging the material distribution to approach either 0 (void) or 1 (solid) [161]. This penalization is achieved through a power-law relationship between the material density and stiffness, which helps in reducing the presence of gray elements (intermediate densities) in the final design. The SIMP method is known for its simplicity and computational efficiency, making it a popular choice for various engineering applications [165].

One of the key advantages of the SIMP method is its ability to generate clear and distinct topologies. By penalizing intermediate densities, SIMP ensures that the final design is predominantly composed of solid and void regions, which is often more manufacturable and easier to interpret [161]. This is particularly useful in industrial applications where clear boundaries between materials and voids are crucial. For instance, in the topology optimization of functionally graded materials (FGMs), the SIMP method has been successfully used to maximize fracture resistance by optimizing the distribution of the second phase [165].

However, the SIMP method is not without its limitations. One of the main drawbacks is the potential for numerical instabilities, such as checkerboard patterns and mesh dependency, which can lead to less smooth and less meaningful topologies. These issues can be addressed through the use of filtering schemes and other numerical stabilization techniques, as demonstrated in several studies [161,166].

The SKB method, on the other hand, is a variant of the Bi-directional Evolutionary Structural Optimization (BESO) method. Unlike the hard-kill approach, which abruptly removes elements from the design domain, the soft-kill method gradually reduces the stiffness of elements that are deemed unnecessary. This gradual approach enables smoother transitions between solid and void regions, resulting in more realistic and smooth topologies [154,164].

The SKB method is particularly useful in cases where smooth transitions are critical, such as in the design of structures with multiple material phases or in applications where manufacturability is a key concern. The soft-kill approach also helps to avoid the convergence issues often associated with hard-kill methods, making it more robust for large-scale structures [154,164]. For example, in the context of multimaterial topology optimization, the SKB method has been shown to effectively handle the distribution of multiple material phases while maintaining smooth boundaries [163].

Despite its advantages, the SKB method can be computationally more expensive than the SIMP method, especially for large-scale structures. This is because the soft-kill approach requires iterative updates to the material properties, which can increase the computational burden. However, recent advancements in computational techniques, such as GPU-based algorithms, have helped mitigate these issues, making the SKB method more feasible for practical applications [162].

Given the strengths and weaknesses of each method, researchers have explored hybrid approaches that combine the SIMP and SKB methods. These hybrid methods aim to leverage the computational efficiency and clarity of SIMP while incorporating the smoothness and robustness of the SKB method. For example, one study proposed a hybrid BESO method that combines the soft-kill and hard-kill approaches for large-scale structures, achieving reduced execution time, decreased memory consumption, and improved FEA convergence [162].

Similarly, other studies have integrated SIMP with evolutionary algorithms, such as genetic algorithms (GAs), to enhance the optimization process. These hybrid methods have shown promising results, with improved convergence speed and better optimization effects compared to traditional methods [154,164].

While the focus on topology optimization presents significant advantages in material design, it is essential to consider the potential for failure due to defects and high void fractions. Balancing optimization with structural integrity remains a critical challenge in the field of metamaterials. For a metamaterial sample that consists of semi-interactively different materials, the void fraction after topology optimization is probably very high, given the mechanical properties of the sample system [167]. There could be one or more cracking locations that, combined with high stresses, lead to the splitting failure of the material. In this case, if the desired mechanical properties of the metamaterial sample are described at the micro-level, it is possible to realize the actual design requirements. In the process of topology optimization, to complete the solution of the topology, the volume elements connected to the kth need to be determined [167]. Topology optimization enhances the mechanical properties of metamaterials by allowing for tailored designs that can achieve desired stiffness and energy absorption characteristics [168].

The optimization process can lead to the emergence of complex structures that exhibit unique mechanical behaviors, such as increased stiffness and improved energy absorption under deformation [168].

The presence of topological defects can significantly influence the mechanical response of metamaterials, potentially leading to failure modes like cracking under high stress [169].

Understanding these defects is essential, as they can alter the material's properties and overall performance, emphasizing the need for careful design and optimization [169].

The design of functionally graded metamaterials allows for varying mechanical properties across the material, which can be optimized for specific applications, enhancing performance and durability [151].

Advanced optimization techniques, including genetic algorithms, can facilitate the design of energy-absorbing structures that meet specific stress–strain requirements, further mitigating failure risks [170].

Integrating topology optimization into metamaterial design not only improves performance but also facilitates the creation of structures with complex geometries that cannot be achieved using traditional methods. This integration utilizes advanced computational techniques, enabling the design of structures that optimize material distribution to achieve desired properties, such as sound absorption and mechanical strength [171]. Topology optimization improves the acoustic performance of metamaterials, as demonstrated in studies where optimized structures achieved superior sound absorption capabilities compared to conventional designs [172].

The mechanical properties of metamaterials can be tailored through optimization, leading to lightweight yet strong structures, which are crucial in applications like the aerospace and automotive industries [173].

The use of additive manufacturing in conjunction with topology optimization allows for the fabrication of intricate designs that traditional manufacturing methods cannot produce. This includes self-supporting structures that eliminate the need for additional support during production [173].

Anisotropic properties of materials can be effectively incorporated into the design process, further enhancing the complexity and functionality of the resulting metamaterials [173].

While the benefits of integrating topology optimization into metamaterial design are substantial, challenges remain, including the computational cost and time associated with the optimization process. These factors can limit the practical application of such advanced techniques in real-world scenarios [174]. Topology optimization has proven valuable in designing metamaterials with specific mechanical properties, such as negative Poisson's ratios. Qin et al. introduced an advanced topology optimization method that enables the design of lightweight metamaterials with tuned Poisson ratios and demonstrated the method's ability to automatically determine optimal configurations rather than relying on predefined shapes [175]. This capability was further supported by the work of Vogiatzis et al., who used a refined level set method to optimize negative-Poisson-ratio metamaterials (including both single- and multiple-material), highlighting the flexibility and effectiveness of topology optimization in achieving desired material distributions [172].

Furthermore, the applicability of topology optimization extends beyond mechanical properties to thermal and acoustic metamaterials. Sha et al. explored the potential of extending 2D thermal metamaterials to 3D configurations through tailored mathematical models, emphasizing the role of topology optimization in improving thermal management capabilities [176]. Similarly, Mendez et al. examined computational material design for acoustic cloaking, where the introduction of topological derivatives enabled the design of metamaterials across multiple length scales, demonstrating the versatility of topology optimization in metamaterial applications [177].

Integrating topology optimization with other design methods, such as shape optimization, has proven to offer significant advantages. Stankiewicz's research on auxetic piezoelectric metamaterials showed that combining topology and shape optimization can leverage the strengths of both approaches and lead to improved design results [178]. This hybrid approach was further validated by Han et al., who demonstrated the practicality of combining topology and shape optimization methods for bifunctional metamaterials by the obtained experimental results that closely matched the optimization predictions [179]. Zeng et al. highlighted the need to consider large geometric deformations in the topology optimization of energy-absorbing structures, suggesting that traditional small deformation assumptions may be insufficient for advanced metamaterial designs [170]. This finding underscores the importance of adapting and refining optimization techniques to address the unique challenges posed by the dynamic and complex behavior exhibited by metamaterials.

3. Metamaterial Fabrication Techniques

Various fabrication methods are required to precisely control the structure at different levels (micro, nano) and to obtain materials with unique properties that differ from those of naturally occurring materials. Metamaterials can be fabricated using both top-down and bottom-up approaches (Figure 4).

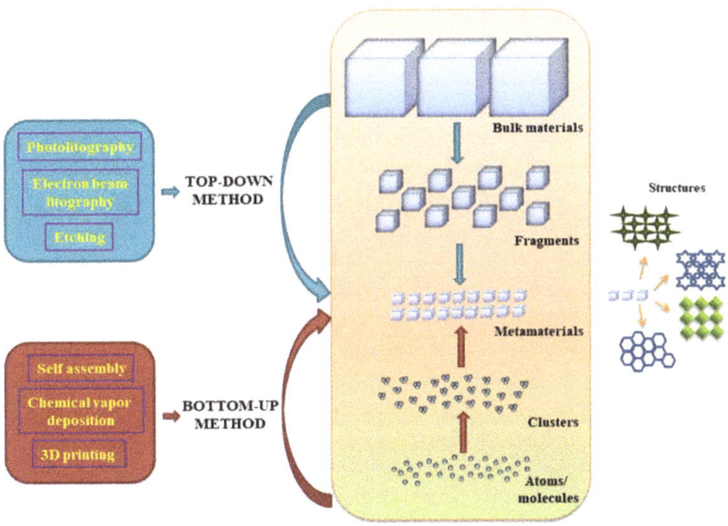

Figure 4. A visual overview of top-down and bottom-up approaches, along with their respective principles.

The top-down technique (applicable for large-scale manufacturing) involves the fabrication of metamaterials from bulk to micro/nanoscale, while bottom-up approaches facilitate the growth and self-assembly of atoms and molecules to produce micro/nanoscale structures.

Top-down techniques, such as photolithography, electron beam lithography, and etching processes, imply a gradual process of reducing and constructing the metamaterial structure, from macroscopic dimensions to precise nanoscale features [97,159,180]. These methods, however, can be limited by scalability and cost [71,180].

Photolithography, a microfabrication technique, is used to manipulate single- and multilayer metamaterials designed to operate at terahertz frequencies. This method is based on the photopolymerization of monomers and enables the achievement of high-resolution subwavelength structures [1]. Light-sensitive photoresists, which undergo chemical changes when exposed to light, enable patterning at the nanoscale. Photolithography, a low-cost method with high efficiency, as described in ref. [181], is restricted due to limited diffraction. Therefore, derived techniques (from traditional photolithography), such as shift mask lithography [182] deep ultraviolet immersion lithography [183], and electron beam lithography [184], have been proposed to overcome this restriction. Despite these advancements, photolithography still faces challenges in effective manufacturing and must be adapted to enhance functionality and precision. Electron beam lithography (EBL) represents a powerful tool for the rapid prototyping of metasurfaces [185]. This method involves generating a pattern by focusing a Gaussian electron beam onto a resistant material to break chemical bonds. Despite its long time processing and high operating costs [181,185], EBL is used for manufacturing large-area surfaces and provides high-resolution patterning for nanoscale structures [181]. It builds upon the principles of traditional photolithography, incorporating enhancements to processing and patterning while reducing the diffraction limitation. Etching processes have evolved over the past years, leading to the development of various technologies, including chemical etching [171] (involving the reaction between species), physical etching (incorporating removal of the material) [186], and ion beam etching [182]. The substrate etching approach is suitable for metamaterials as it improves sensitivity and reduces the effective refractive index [183]. Regarding the ion beam etching technique, an ion beam interacts with the material's surface without altering the internal structure, which can contribute to the fabrication of metamaterials with complex geometries [1,182]. However, this method is often limited to several structures, and the used equipment is expensive [1].

Bottom-up approaches, such as self-assembly [187], chemical vapor deposition (CVD) [188], and 3D printing [189], ensure promising alternatives for producing complex structures at lower costs [190–192]. The self-assembly process involves the organization of individual components into ordered or unconventional patterns or structures [193,194]. This process can occur with either active materials that interact according to predetermined programs or passive materials that interact in line with intermolecular forces, geometry, or surface chemistry [189]. This method is used to synthesize different types of nanoparticles for surface functionalization, along with a variety of metamaterials, with potential applications in the optical field, electromagnetic domain, or other fields [189,194,195]. The self-assembly process holds significant potential for synthesizing large-dimension metamaterials with homogenous properties, making them ideal substrates for energy materials [193,196]. It is also an effective approach for assembling plasmonic metamaterials and fabricating sensor-based metamaterials. However, the self-assembly technique is usually limited regarding system phase distribution and the templates used. Single-phase system metamaterials are usually determined, while multi-phase systems or complex systems have not been widely explored [196]. As for templates, they are also limited to certain types of particles [197]. An alternative method for fabricating metamaterials is chemical vapor deposition (CVD), often combined with physical vapor deposition (PVD) to create complex and controlled structures over large uniform areas [198]. Compared to other techniques, CVD is remarkable for its versatility in depositing a wide range of materials for various fields (e.g., material science [199], medicine [200] or engineering), thus achieving conformal layers with high production efficiency and low-cost fabrication. Unlike PVD, which uses the vaporization of solid precursors for deposition, the CVD process deposits solid materials onto a substrate by introducing gaseous precursors, often in combination with carrier gases, which react to form thin films or coatings [201]. The disadvantages of this method include the production of chemical compounds and limited low-pressure and high-temperature depositions [199,202]. Three-dimensional printing or additive manufacturing (AM) expands the possibilities of creating metamaterials with unique electrical and transmission properties that cannot be obtained by conventional approaches [203]. Advanced metamaterials can be fabricated using AM by controlling the structure of patterns and topology, offering high design freedom for achieving negative stiffness and a negative Poisson's ratio [204]. Three-dimensional printing technologies include different methods such as inkjet printing [203] selective laser sintering (SLS) [204], fused deposition modeling (FDM) [205], and stereolithography (SLA) [206], each with specific advantages and disadvantages but following similar fundamental working principles while providing a fast, convenient, and efficient route for the fabrication of metamaterials with a low production cost [74]. AM has certain limitations. For example, the extrusion temperature is limited, resulting in poor surface quality, and, consequently, the final product exhibits weak mechanical properties [207].

Such techniques facilitate the production of intricate geometries, including multimaterial and functionally graded metamaterials, which are essential for applications requiring tailored thermal, mechanical, or optical properties [208].

Thus, 3D-printed mechanical metamaterials with unique properties, such as negative Poisson's ratios, have been developed for lightweight aerospace components and energy-absorbing structures [208]. Furthermore, 4D printing, which incorporates stimuli-responsive materials, allows for the creation of dynamic metamaterials that adapt their properties in real time to environmental changes [209].

In summary, top-down methods offer high precision but can be expensive and time-consuming, while bottom-up methods, particularly additive manufacturing, allow greater design flexibility and scalability [71,190]. The choice of a manufacturing technique often depends on the specific application and required material properties [71]. Table 1 summarizes the main metamaterial fabrication techniques and highlights the advantages and disadvantages of each method.

Table 1. Main fabrication techniques of metamaterials.

Approach	Technique	Advantages	Disadvantages	References
Top-down	Photolithography	High-resolution structures, low-cost	Restricted due to limit diffraction, low effective manufacturing	[1,181,210]
	Electron beam lithography	High resolution pattering for nanoscale structures, rapid prototyping	High operating costs, time intensive for large areas	[181,185]
	Etching	High sensitivity, low refractive index	Limited structure, high-cost equipment	[1,182,183]
Bottom-up	Self-assembly	Low-cost, suitable for different types of nanoparticles, homogenous structures	Restricted template types, limited by phase distribution	[193–196,210]
	Chemical vapor deposition	Complex and controlled structure, high production efficiency, conformal layer deposition, low-cost fabrication	Low pressure and high temperature limitations, production of chemical compounds waste	[198,211]
	3D printing	Design freedom, suitable for developing advanced metamaterials, low-cost production	Limited extrusion temperature, weak mechanical properties, poor quality surface	[203–205]

Furthermore, the design and fabrication of metamaterials are inherently tailored to their intended functions, with each application demanding specific structural and material properties [212]. Consequently, metamaterial fabrication techniques must be carefully selected based on the specific functional requirements of each application. For instance, electromagnetic applications, such as cloaking devices, demand nanoscale precision fabrication, achievable through EBL or photolithography, to precisely control permittivity and permeability [181,185] AM offers a distinct advantage in the fabrication of metamaterials for energy absorption, enabling the creation of complex auxetic geometries [119,204]. Similarly, thermal management applications require the CVD technique for controlled thermal conductivity gradients [213,214], while acoustic metamaterials utilize self-assembly techniques to achieve specific resonant properties that manipulate sound waves [189,194].

Fundamentally, the choice of a fabrication method, whether top-down (e.g., etching for optical metasurfaces [182,183]) or bottom-up (e.g., 3D printing for tissue scaffolds [215,216]), is guided by the intended metamaterial function and operational requirements, including scale and desired anisotropy [190]. The implementation of a function-specific approach ensures optimal performance by addressing the distinct demands of each application, ranging from medical biosensors requiring nanoscale precision, through EBL [148,217,218], to large-scale structural applications where scalability is paramount, through AM [154,173]. Thus, the interplay between function-driven design and targeted fabrication is critical to realizing the unique properties of metamaterials across diverse applications [14].

4. Metamaterials in Medicine

Metamaterials present unique properties and features that have found a lot of applications in various domains [49], with a remarkable continuous development of metamaterials in optical [219], imaging [220], and medical fields. One of the most attractive applications in optics and imaging based on metamaterials is represented by superlenses [49,220]. A superlens is a metamaterial-based optical device that manipulates and controls the near-field interactions and enhances the evanescent waves typically associated with nanoscale sources. By exciting magnetic and electric surface modes, it couples with these near fields and focuses them on the opposite side, enabling the imaging of nanoscale objects that traditional lenses cannot resolve [213,221]. Due to the naturally occurring negative permittivity ($\varepsilon < 0$) in materials, such as silver and other metals at visible wavelengths, a thin metallic film can function as an optical superlens [222]. Other types of lenses, including beam steering lenses and antenna lenses, have been proposed, analyzed, and realized using the gradient refractive index of metamaterials, which can convert the generated wave into a quasi-plane wave [149]. Recent developments have shown a great interest in metamaterials in medicine and bio/medicine. Driven by the need for diagnosis and continuous monitoring of vital parameters, even outside the clinical environment, metamaterials with various structures tailored for different purposes have found applications in medicine [201,223]. Excellent results have been achieved using, for instance, negative-index media that refract light in the opposite direction of conventional materials, chiral materials that change the polarization state of light, and thin films with strong dispersion, which work as a resonant atomic system with electromagnetically induced transparency. These results produced a boost for the use of novel metamaterials in the biomedical field [223]. Metamaterials have been developed for different medical applications, encompassing mechanical metamaterials, auxetic metamaterials [45,224], optical metamaterials [225], and acoustic metamaterials [46]. For instance, mechanical metamaterials with zero or negative Poisson ratios and unique constitutive behaviors offer innovative solutions for tissue engineering and functional implants [203,226]. These materials possess specific properties suitable for the medical domain, including a low modulus of elasticity, high and negative stiffness, high strength, and ultra-low weight. A prosthesis or implant fabricated from a metamaterial with a modulus of elasticity close to bone, while ensuring toughness and high stiffness, will behave like a natural bone. Moreover, by modulating their structure, specific properties can be obtained in specific areas to meet the needs of particular medical applications. Thus, mechanical metamaterials will behave more naturally because they can overcome the challenge of preventing bone sclerosis and resorption by creating favorable mechanical junctions and angles [201,226]. Auxetic metamaterials, another type with excellent mechanical properties, exhibit the uncommon properties of lateral expansion when stretched and densification when compressed. These materials, comprising a special group of metamaterials with varied geometrical structures, shapes, and auxetic foams, have found application in biomedical devices involving prostheses, tissue engineering, orthopedics, in vitro devices, and advanced clinical devices [45]. For example, auxetic films show huge potential in the biomedical field as they can control the normal function of different organs of the body, providing the necessary mechanical properties for a wide range of biomaterials or medical devices, such as bone screws, intervertebral discs, or skull implants [224,227]. Optical metamaterials in the medical field are commonly used as biosensors. These devices are highlighted for several advantages, providing a fast method to identify specific objects and to detect different chemical compounds in real-time and cost-effectively. They operate across a broad spectrum, from visible to infrared (IR) light, and are non-invasive with high sensitivity [228]. Nevertheless, acoustic metamaterials are widely used in medicine for drug delivery [229] and noise imaging reduction [230] in signal acquisition equipment. Their artificial design structures that break through sound waves are capable of manipulating and controlling sound waves at different wavelength scales.

Several studies have explored the metamaterials previously mentioned, each offering unique properties that enhance their potential applications in the medical/biomedical field. Medical antennas represent a crucial part of medical systems that facilitate communications between the patient and the monitoring stations [201,225,231]. In recent years, the demand for antennas providing wireless telemetry systems in medicine has grown significantly, driven by the need for early diagnosis and continuous monitoring of different parameters [225]. Designing these metamaterial-based antennas often involves periodic SRRs, which are known for their negative permeability and dielectric constant of less than 1, in conjunction with metallic posts [225,231]. Saban's research focused on developing an antenna using printed metamaterials, thus demonstrating a greater efficiency compared to a regular antenna. By incorporating SRRs and metallic strips, the antenna achieved an improved bandwidth, with a Voltage Standing Wave Ratio (VSWR) better than 2:3:1, shifting the resonant frequency on the human body by 3% [231]. Saban continued his research by optimizing small antennas for integration into communication systems for receiving and transmitting signals. This resulted in an improved bandwidth, from 50% up to 100%, while the resonant frequency was enhanced by up to 5% when the antenna was placed on the human body [225]. Hosseinzadeh et al. mentioned that one of the most significant applications of medical antennas is the wireless endoscope, which is a capsule-shaped system used for gastrointestinal monitoring or treatment [201]. Further, Yue et al. designed a miniaturized and ultra-thin midfield wireless energy transmission antenna utilizing metamaterial structures [218]. Metamaterials have also found promising applications in radio bands used for medical purposes. Raghavan and Rajeshkumar explored the use of metamaterial antennas for Industrial, Scientific, and Medical (ISM) radio bands, a part of the radio spectrum reserved for the use of radio frequency. The designed antenna was suitable for dual-frequency bands within the MHz range, being an excellent candidate for medical applications within the ISM spectrum [223]. Cancer is characterized by the uncontrolled proliferation of malignant cells, which can grow in nearly all body parts. Thus, early detection and localization are critical for effective treatment. In this respect, Vafapour et al. studied a metamaterial sensor antenna for detecting breast cancer cell lines. They placed an SRR atop a substrate, while a third iteration of the Minkowski fractal curve was used as a Defected Ground Structure, in the ground plane. This design proved effective in investigating the electrical characteristics of the cancer cell lines, and the simulated results determined resonant frequencies corresponding to different constants of breast cancer cell lines [148,217].

The Minkowski fractal antenna, designed for breast cancer detection, demonstrated resonant frequencies of 2.35, 2.42, and 2.52 GHz when tested with various breast cancer cell lines, including MDA-MB-231 and MCF-7 [217].

A novel graphene buzzle metamaterial refractometer achieved high sensitivity, with an average sensitivity of 495 GHz/RIU, effectively distinguishing between different cancer cell types based on their dielectric properties [214].

Hosseinzadeh et al. also highlighted the potential for developing microwave devices combined with metamaterial structures, which can lead to a cost-effective device, more specifically, a biosensor with high precision for localizing malignant cells [232]. Hamza et al.'s study aimed to develop an innovative multiband biosensor design for the early detection of cervical cancer, operating across the frequency range of 0 to 6 THz.

The development of metamaterial-based biosensors represents a significant technological advancement in early cancer detection, offering unprecedented sensitivity and real-time monitoring capabilities [109,148] through the engineered enhancement of light–matter interactions via subwavelength structures [92]. These sensors demonstrate a detection sensitivity 100–1000 times greater than conventional electrochemical or optical biosensors [35], with a designed split-ring resonator (SRR) that can achieve attomolar (10^{-18} M) detection limits for cancer biomarkers [148].

The sensor architecture, consisting of a patterned aluminum layer on a polyimide substrate, is carefully optimized to achieve near-perfect absorption. Full-wave electromagnetic simulations were crucial in the design process, enabling a thorough analysis of the sensor's absorption characteristics and the impact of material properties [223].

The corona-shaped metamaterial biosensor exhibited high sensitivity and selectivity, confirming its capability to detect cancer markers in real time [233]. Yang et al. confirmed that mechanical metamaterials improve the sensitivity of sensors, which is crucial for monitoring physiological changes in real time. These materials enhance the sensitivity of biomolecule detection, facilitating precise diagnostics through their high penetrability in biological tissues [189].

These sensors also facilitate real-time monitoring, with wireless designs demonstrating continuous tracking of tumor progression through frequency shifts of 183.2 GHz per growth stage [233] and microfluidic-integrated designs achieving rapid response times of less than 5 min, exceeding traditional methods performance like ELISA [35].

Compared to conventional biosensors, metamaterial biosensors offer significantly lower detection limits (attomolar vs. picomolar), faster response times (minutes vs. hours), greater multiplexing capabilities (5–10 biomarkers vs. 1–2), smaller size (sub-mm vs. cm-scale), and reduced cost per test (USD 10–50 vs. USD 100–500) [92,216]. These attributes render them a promising alternative for early and accurate cancer diagnosis [109].

A flexible terahertz metamaterial biosensor recorded a maximum resonance peak frequency shift of 183.2 GHz when monitoring breast cancer cell growth, indicating its potential for non-destructive detection [233].

In the field of tissue engineering, mechanical metamaterials are the most suitable type of medical metamaterial due to their ability to manufacture scaffolds that support cell attachment and direct their functions.

Mechanical metamaterials offer a revolutionary approach to tissue engineering by providing unprecedented control over mechanical properties through their designed internal architectures. This enables the creation of scaffolds that more accurately mimic the complex mechanical environment of native tissues compared to traditional scaffold materials [46,234]. These metamaterials exhibit enhanced mechanical compatibility by precisely tuning the elastic modulus, Poisson's ratio, and energy absorption [140,235]. Furthermore, they offer improved biocompatibility through structural design that promotes cell attachment and mechanotransduction, controlled degradation rates, and interconnected pore networks for vascularization [189,236]. Their capacity to achieve functional gradation, characterized by spatially varying and dynamically responsive properties, addresses the limitations of homogeneous traditional scaffolds [237]. While fabrication complexity and scalability are ongoing concerns [234], the current applications in bone, cartilage, and skin regeneration [123,238], along with future directions in bioactive coatings, biodegradable materials, and sensing capabilities [239] highlight the significant potential of mechanical metamaterials in advancing tissue engineering.

Techniques like bioprinting allow for the creation of bioactive bone implants and artificial tissues, utilizing diverse materials such as biodegradable polymers and metals [240].

Structures inspired by nature, such as triply periodic minimal surfaces, enhance mechanical compatibility and functionality in tissue scaffolds [234].

Mahmud et al. demonstrated that metamaterials optimize wireless power transfer systems, addressing challenges like tissue absorption and coupling efficiency, thus enabling longer-lasting implantable medical devices [241].

Moreover, they can be used to fabricate advanced scaffolds, expand the forms of mechanical signals, and enhance the cellular response towards tissue regeneration [140,189]. Dogan et al. conducted a comprehensive review of various tissues (e.g., cartilage, bone, skin, and vascularized tissues) that can be regenerated using mechanical metamaterials, with a particular focus on those fabricated by 3D printing, and suggested various guidelines for fabricating metamaterials in the field of medicine [58]. Three-dimensional-formed hydrogels are often used as a scaffold for regenerating tissue [189,215].

Hydrogel-based scaffolds offer unique physicochemical properties that closely resemble the natural extracellular matrix (ECM) [189,215]. Their superior biocompatibility and bioactivity distinguish them from alternative materials, such as ceramics, metals, and synthetic polymers. Notably, hydrogels derived from natural polymers (e.g., alginate, hyaluronic acid), exhibit excellent biocompatibility, minimizing immunogenic responses and promoting cellular adhesion, proliferation, and differentiation [215,236]. Their high water content (80–99%) replicates the native tissue environment, facilitating efficient nutrient diffusion and waste removal, which is crucial for cell survival [189].

Furthermore, hydrogels offer tunable mechanical and degradation properties. Unlike rigid ceramic- or metal-based scaffolds, they can be engineered to match the mechanical properties of soft tissues (e.g., cartilage, skin) through crosslinking density and polymer composition [215].

Hong et al. [242] emphasized the importance of mechanical stability against the physical impact of hydrogels for effective tissue regeneration. They explored how the mechanical properties of hydrogels could be regulated by adjusting the molecular weight, thus affecting the mobility of polymer chains. According to their research, the tunability allows for the creation of a hydrogel-based cell matrix with biodegradable support, playing a crucial role in the stem cell proliferation process [215]. Metamaterials also hold promise for bone regeneration, where they can be integrated into implants to promote bone tissue growth once they interact with the tissue.

Degradation kinetics can also be modulated to synchronize with tissue regeneration, preventing premature structural failure [236]. The porous, hydrated structure of hydrogels further facilitates enhanced drug and growth factor delivery, enabling the controlled release of bioactive molecules (e.g., growth factors) and reducing systemic side effects [189,243].

Bone tissue, a hierarchical structure with both mineralized and non-mineralized components, requires extreme compressive strength and superior toughness. The use of light-assisted 3D technology could provide a wide range of options for creating advanced structures to meet these requirements [189]. Lecina-Tejero et al. examined different studies on auxetic materials for skin regeneration and revealed that human skin is a complex and large organ with three heterogeneous layers, populated by different cell types with different functional, mechanical, and biological characteristics [189,238]. These layers, the epidermis, the dermis, and the hypodermis, are also different in terms of structure or composition. For example, the dermis consists of a collagen network with elastic fibers that provide tissue toughness and hyperelasticity [58]. N. Li et al. proposed a novel electronic skin (e-skin) that combines the unique properties of mechanical metamaterials with self-powering capabilities, multimodal fusion perception, and shape memory reconfigurability. The e-skin incorporates multimodal sensory capabilities, effectively replicating a biomimetic sensory system's auditory, tactile, and visual perception [244]. In another study, Kirillova et al. presented a novel 4D biofabrication method capable of producing hollow self-folding hydrogel tubes, employing alginate and hyaluronic acid hydrogels as representative models. This technique combines polymer–cell bioinks in a simultaneous printing process, creating shape-morphing biopolymer hydrogels that facilitate the precise fabrication of these advanced structures, with applications in tissue engineering and regenerative medicine [236]. Zeenat et al. summarized major bioprinting techniques, highlighting two main approaches for forming vasculature within 3D printed structures. The first one involves the release of angiogenic factors to stimulate the growth of vasculature, while the second focuses on the direct printing of the vascular network to target the tissue cell. Both approaches provide precision and control over the internal and external structure of 3D-printed constructs [243]. Medical imaging techniques represent non-invasive methods to diagnose or treat different medical conditions. Kasban et al. conducted a detailed comparison of medical imaging techniques, covering their current concepts, risks, challenges, advantages, and applications. Among these, they considered Magnetic Resonance Imaging (MRI), a radiation-free imaging technique used to visualize the morphological changes in diseased tissues [232]. Wang et al. discussed future challenges and opportunities for nanomaterials as MRI agents, emphasizing key applications such as tissue necrosis, clinical diagnosis, and local ischemia [33].

MRI image generation is predicated on the detection of spatial variations in the phase and frequency of radiofrequency (RF) waves, which are absorbed and subsequently emitted by the nuclear spins of the imaged object. Standard MRI detectors, utilizing conventional coils, operate within the near-field regime of the RF magnetic field. Metamaterials offer the potential for improved coil performance by enabling precise control over this magnetic near field [245].

B. Li et al. further explored the potential of MRI, suggesting that the recent advancements in metamaterials could significantly enhance imaging capabilities. They proved that metamaterial-based lenses can enhance the signal-to-noise ratio (SNR) even during accelerating scans because of their capabilities to enable interaction with radiofrequency (RF) and to amplify it [246]. Moreover, integrating nano-metamaterials can contribute to MRI contrast agents due to their complex microhierarchical structure [33].

In contrast to the resonant operation of conventional MR coils, the $\mu_r = -1$ metamaterial lens, based on split-ring resonators, functions at a frequency distinct from the individual resonance of its constituent elements. This operational deviation, arising from the collective response of the split rings, which yields the desired effective permeability, contributes to a reduction in both losses and noise [247].

Several metamaterial designs have been developed to boost the SNR. For instance, Duan et al. reported a magnetic metamaterial composed of an array of unit cells featuring metallic helices to improve the SNR by enabling field enhancement around a specific resonant mode. Their study confirmed the suitability of magnetic metamaterials for SNR improvement, increasing image resolution and scan efficiency. Also, it was concluded that the resonant mode, when applied to MRI, could be excited by the radiofrequency magnetic field and highly boost the SNR [248]. For preclinical imaging, a dual-nuclei radiofrequency coil was developed and assessed by Hurshkainen et al., leveraging the resonant excitation of eigenmodes within a pair of multimode wire metamaterial-inspired resonators. Numerical and experimental analyses demonstrated that the strategic selection of excited eigenmodes enables precise control over the penetration depth within a subject, facilitating either whole-body excitation or targeted surface excitation in small animals. For multi-nuclei MRI applications, the proposed coil design offers compatibility through the tunable nature of its short and long-wire resonators. Larmor frequency tuning is facilitated by manipulating resonator length and patch parameters, with the patches providing distributed load capacitance for the short-wire resonator [249]. Overall, recent advancements in metamaterials have significantly enhanced MRI performance, offering higher-resolution images with improved SNRs.

Furthermore, Jiang et al. conducted extensive research on stretchable strain sensors. In their two reviews, they emphasized that strain sensors that convert mechanical stimuli into electrical signals [250] or optical signals [251] play an important role in medical applications. They concluded that the sensitivity of stretchable strain sensors is an important parameter that can be improved by using specific metamaterials. The integration of auxetic metamaterials into strain sensor design has proven to increase their sensitivity by reducing the structural Poisson's ratio and strain concentration [251]. In addition, the incorporation of wireless force sensing within intelligent implantable devices facilitates real-time, in situ monitoring of mechanical stress, thereby enabling dynamic parametric modulation for optimized functional performance. Luo et al. developed proof-of-concept metamaterial orthopedic implants designed to extract energy from natural body movements. The resulting electrical signal was utilized for direct, wireless, and electronics-free transmission of sensor data, eliminating the need for traditional electronic components. These implants allow for real-time wireless communication over moderate distances, with power outputs measured in the picowatt (pW) range [252]. Various studies have explored additional applications of metamaterials in medicine, including microwave hyperthermia [253,254], radiation shields [254], and hearing aids [160]. An overview of these applications can be found in Table 2.

Table 2. Various applications of metamaterials in medicine.

Application	Specific Use	Benefits	Metamaterial Properties	References
Optical antennas	Provides effective communication in medical systems	High efficiency	Negative permeability, subunit dielectric constant	[26,223,225,231]
Cancer detection	Detects early signs of cancer by localizing malignant cells with biosensors	High precision, cost-effective	High sensitivity, negative values of relative permittivity, electromagnetic properties	[148,201,223]
Tissue engineering	Scaffolds for regenerating different tissues such as bone, skin, cartilage, or vascular	Enhanced stem cell proliferation, enhanced biocompatibility, natural-tissue-like behavior	Mechanical stability, high strength and toughness, precise control over structure	[104,189,215,243]

Table 2. *Cont.*

Application	Specific Use	Benefits	Metamaterial Properties	References
Medical imaging	MRI enhancement for accelerated scans and better SNRs	High-resolution images, simple installation	Magnetic, resonant, and dielectric properties	[201,232,248]
Strain sensors	Monitories healing of fractured bones, mechanical stimuli detection	Enhanced sensitivity, non-invasive	Auxetic properties, distinctive resonance frequency, reduced Poisson's ratio	[201,250]
Microwave hyperthermia	Superficial cancer treatment in specific locations by focusing resolution	Environmentally friendly, non-invasive, enhanced distribution field	Electromagnetic convergence, thermal and heat properties	[253,255,256]
Radiation shields	Protection against extreme electromagnetic radiation of medical equipment	High indicators of transparency, enhanced performance of the system	Absorption properties, tunable and low-profile features	[35,254,257]
Hearing aids	Rehabilitates higher cortical functions	High bandwidth, miniaturization, low mutual coupling	Filtration and dielectrics properties, resistance to mechanical properties	[160,240,258]

5. Conclusions, Challenges, and Future Directions

This review provides an overview of the latest advancements in the design and processing of the key controlled-structure metamaterials, underscoring their potential to enhance performance in various fields. The discussion highlights the challenges faced by metamaterials in achieving high performance and efficiency, the complexity of developing novel and unique designs, and the potential future directions for the field. While significant progress has been made in characterizing and optimizing metamaterials and metadevices, with notable successes over time, we have concluded that further improvement in this area is necessary to address the current challenges and overcome limitations. The various fabrication techniques available for producing metamaterials offer a wide range of approaches and different perspectives for the development of innovative devices. However, despite their promising potential, their performance and characteristics have some limitations and challenges. Metamaterials are frequency-dependent on the applied electromagnetic wave, which can impact their efficiency in specific applications. In addition, the negative permittivity and permeability of certain metamaterials often lead to material and energy losses due to high absorption and scattering. The fundamental limitations of these manufacturing approaches make it challenging to determine the ideal method, one that not only meets the stringent efficiency requirements but also delivers good performance across various parameters for different aspects. Costly equipment, expensive operating systems, and time-consuming process maintenance are among the primary hurdles in producing metamaterials. Although metamaterials can achieve complex structures and geometries compared to traditional materials, their validation requires advanced computer simulations, including numerical methods and experimental techniques, which are often time-consuming and require specialized expertise.

The integration of practical fabrication techniques for efficient and rapid prototyping of complex metamaterials design with quality control methods can ensure the uniformity and reproducibility of metamaterials. Advanced techniques, such as EBL, photolithography, or AM, enable the achievement of complex nanoscale patterns, with quality and uniformity assessed through characterization techniques like scanning electron microscopy (SEM) or atomic force microscopy (AFM). Nevertheless, scalability and cost-effectiveness remain pivotal factors for the commercial viability of metamaterials. AM technology offers a pathway to the large-area production of metamaterials without compromising the quality of the final product. On the other hand, techniques such as photolithography or self-assembly propose low manufacturing costs, prioritizing high production rates over expensive devices with complex parts that require costly maintenance. Integrating these methods into the manufacturing process of metamaterials ensures good scalability at an affordable price, facilitating their use in various fields, including telecommunications, electronics, and medicine.

This comprehensive examination of the current status of metamaterials reveals their ability to exhibit unique properties through complex design. We have explored the historical evolution of metamaterials, their theoretical principles, various manufacturing methods, and their applications, particularly in the medical field. Future developments in manufacturing efficiency, as well as advancements in computational tools, will create new opportunities for the design of innovative and specialized metamaterials that will surpass current limitations and play a crucial role in various areas in the years to come.

Author Contributions: Conceptualization, G.F.P.-P. and A.I.V.; methodology, A.T.M.; validation, G.F.P.-P. and A.I.V.; writing—original draft preparation, A.T.M.; writing—review and editing, G.F.P.-P. and A.I.V.; visualization, G.F.P.-P. and A.I.V.; supervision, G.F.P.-P. and A.I.V.; project administration, G.F.P.-P. and A.I.V.; funding acquisition, G.F.P.-P. and A.I.V. All authors have read and agreed to the published version of the manuscript.

Funding: The authors acknowledge a grant from the Ministry of Research, Innovation and Digitalization, CCCDI—UEFISCDI, project number PN-IV-P2-2.1-TE-2023-0993, within PNCDI IV. This research was also funded by the Romanian Ministry of Research, Innovation and Digitalization, under the Romanian National Nucleu Program LAPLAS VII—contract No. 30N/2023.

Conflicts of Interest: The authors declare no conflict of interest.

Abbreviations

EM	electromagnetic
LH	left-handed materials
BW	backward wave
MNG	mu-negative materials
ENG	epsilon-negative materials
DPS	double-positive materials
THz	terahertz
NIM	negative-index metamaterial
DNG	double-negative
SRR	split-ring resonator
FDTD	finite-difference time-domain
FEM	finite element method
MEMSs	microelectrochemical systems
LSPRs	localized surface plasmon resonances
Poisson's ratio	A measure of the proportional decrease in diameter to increase in length in a stretched material

ADMs	all-dielectric metamaterials
CTE	Coefficient of Thermal Expansion
PESs	permittivity and permeability of electrical substances
HPLs	High-Performance Lenses
SLS	selective laser sintering (often used in 3D printing)
CVD	chemical vapor deposition
3D printing	Refers to various additive manufacturing processes for creating objects layer by layer
LC	liquid crystal
TO	topology optimization
SIMP	Solid Isotropic Material with Penalization
SKB	Soft-Kill Binary
FGMs	functionally graded materials
BESO	Bi-directional Evolutionary Structural Optimization
AC	Alternative Current
DC	Direct Current
NMR	Nuclear Magnetic Resonance (sometimes used in the context of imaging)
CT	Computed Tomography
FPs	Frequency Parameters (related to resonant frequencies in metamaterials)
CRP	Composite Resonant Properties
SCMs	Scalable Composite Materials
EBL	electron beam lithography
PVD	physical vapor deposition
AM	additive manufacturing
SLA	stereolithography
VSWR	Voltage Standing Wave Ratio
MRI	Magnetic Resonance Imaging
SNR	signal-to-noise ratio
ISM	Industrial, Scientific, and Medical
SEM	scanning electron microscopy
AFM	atomic force microscopy

References

1. Ali, A.; Mitra, A.; Aïssa, B. Metamaterials and metasurfaces: A review from the perspectives of materials, mechanisms and advanced metadevices. *Nanomaterials* **2022**, *12*, 1027. [CrossRef] [PubMed]
2. Zhang, X.; Liu, Z. Superlenses to overcome the diffraction limit. *Nat. Mater.* **2008**, *7*, 435–441. [CrossRef]
3. Vukusic, P.; Sambles, J.R. Photonic structures in biology. *Nature* **2003**, *424*, 852–855. [CrossRef]
4. Kinoshita, S.; Yoshioka, S.; Miyazaki, J. Physics of structural colors. *Rep. Prog. Phys.* **2008**, *71*, 076401. [CrossRef]
5. Soukoulis, C.M.; Wegener, M. Past achievements and future challenges in the development of three-dimensional photonic metamaterials. *Nat. Photonics* **2011**, *5*, 523–530. [CrossRef]
6. Suresh Kumar, N.; Naidu, K.C.B.; Banerjee, P.; Anil Babu, T.; Venkata Shiva Reddy, B. A review on metamaterials for device applications. *Crystals* **2021**, *11*, 518. [CrossRef]
7. Shamim, S.; Mohsin, A.S.; Rahman, M.M.; Bhuian, M.B.H. Recent advances in the metamaterial and metasurface-based biosensor in the gigahertz, terahertz, and optical frequency domains. *Heliyon* **2024**, *10*, e33272. [CrossRef]
8. Smith, D.R.; Pendry, J.B.; Wiltshire, M.C. Metamaterials and negative refractive index. *Science* **2004**, *305*, 788–792. [CrossRef]
9. Suzuki, T.; Sekiya, M.; Sato, T.; Takebayashi, Y. Negative refractive index metamaterial with high transmission, low reflection, and low loss in the terahertz waveband. *Opt. Express* **2018**, *26*, 8314–8324. [CrossRef]
10. Lorduy G, H.; Castellanos, L. Negative electrical permittivity in metamaterials for a wire of rectangular cross-sectional: An application to antennas design. *J. Electromagn. Waves Appl.* **2020**, *34*, 1842–1848. [CrossRef]
11. Stuart, H.R. The application of negative permittivity materials and metamaterials in electrically small antennas. In Proceedings of the 2007 IEEE Antennas and Propagation Society International Symposium, Honolulu, HI, USA, 9–15 June 2007; pp. 1148–1151.
12. Chen, J.; Qin, G.; Shi, Y.; Pan, K.; Du, J.; Qiu, J. Negative permittivity and negative magnetic susceptibility of polypyrrole nanorings/carbon nanotubes multi-dimensional metacomposites in the radiowave frequency range. *Org. Electron.* **2022**, *104*, 106470. [CrossRef]

13. Eason, K.; Luk'Yanchuk, B.; Zhou, Y.; Miroshnichenko, A.E.; Kivshar, Y.S. Magnetic microscopy/metrology potential of metamaterials using nanosized spherical particle arrays. In Proceedings of the Smart Nano-Micro Materials and Devices, Melbourne, Australia, 4–7 December 2011; pp. 427–432.
14. Lu, C.; Hsieh, M.; Huang, Z.; Zhang, C.; Lin, Y.; Shen, Q.; Chen, F.; Zhang, L. Architectural design and additive manufacturing of mechanical metamaterials: A review. *Engineering* **2022**, *17*, 44–63. [CrossRef]
15. Veselago, V.G. The electrodynamics of substances with simultaneously negative values of ε and μ. *Phys.-Uspekhi* **1968**, *10*, 509–514. [CrossRef]
16. Zheludev, N.I.; Kivshar, Y.S. From metamaterials to metadevices. *Nat. Mater.* **2012**, *11*, 917–924. [CrossRef]
17. Pendry, J.B. Negative refraction makes a perfect lens. *Phys. Rev. Lett.* **2000**, *85*, 3966. [CrossRef]
18. Eleftheriades, G.V.; Balmain, K.G. *Negative-Refraction Metamaterials: Fundamental Principles and Applications*; John Wiley & Sons: Hoboken, NJ, USA, 2005.
19. Papasimakis, N.; Fedotov, V.A.; Zheludev, N.; Prosvirnin, S. Metamaterial analog of electromagnetically induced transparency. *Phys. Rev. Lett.* **2008**, *101*, 253903. [CrossRef]
20. Kurter, C.; Tassin, P.; Zhang, L.; Koschny, T.; Zhuravel, A.P.; Ustinov, A.V.; Anlage, S.M.; Soukoulis, C.M. Classical Analogue of Electromagnetically Induced Transparency with a Metal-Superconductor Hybrid Metamaterial. *Phys. Rev. Lett.* **2011**, *107*, 043901. [CrossRef]
21. Jin, B.; Wu, J.; Zhang, C.; Jia, X.; Jia, T.; Kang, L.; Chen, J.; Wu, P. Enhanced slow light in superconducting electromagnetically induced transparency metamaterials. *Supercond. Sci. Technol.* **2013**, *26*, 074004. [CrossRef]
22. Schurig, D.; Mock, J.J.; Justice, B.; Cummer, S.A.; Pendry, J.B.; Starr, A.F.; Smith, D.R. Metamaterial electromagnetic cloak at microwave frequencies. *Science* **2006**, *314*, 977–980. [CrossRef]
23. Linden, S.; Enkrich, C.; Dolling, G.; Klein, M.W.; Zhou, J.; Koschny, T.; Soukoulis, C.M.; Burger, S.; Schmidt, F.; Wegener, M. Photonic metamaterials: Magnetism at optical frequencies. *IEEE J. Sel. Top. Quantum Electron.* **2007**, *12*, 1097–1105. [CrossRef]
24. Li, C.; Wu, J.; Jiang, S.; Su, R.; Zhang, C.; Jiang, C.; Zhou, G.; Jin, B.; Kang, L.; Xu, W. Electrical dynamic modulation of THz radiation based on superconducting metamaterials. *Appl. Phys. Lett.* **2017**, *111*, 092601. [CrossRef]
25. Shelby, R.A.; Smith, D.R.; Schultz, S. Experimental verification of a negative index of refraction. *Science* **2001**, *292*, 77–79. [CrossRef]
26. Zheludev, N.I. The road ahead for metamaterials. *Science* **2010**, *328*, 582–583. [CrossRef] [PubMed]
27. Tong, L. Micro/nanofibre optical sensors: Challenges and prospects. *Sensors* **2018**, *18*, 903. [CrossRef]
28. Engheta, N.; Alù, A.; Ziolkowski, R.W.; Erentok, A. Fundamentals of waveguide and antenna applications involving DNG and SNG metamaterials. In *Metamaterials: Physics and Engineering Explorations*; IEEE: New York, NY, USA, 2006; pp. 43–85.
29. Capolino, F. *Theory and Phenomena of Metamaterials*; CRC Press: Boca Raton, FL, USA, 2017.
30. Smith, D.R.; Padilla, W.J.; Vier, D.; Nemat-Nasser, S.C.; Schultz, S. Composite medium with simultaneously negative permeability and permittivity. *Phys. Rev. Lett.* **2000**, *84*, 4184. [CrossRef]
31. Wang, X.; Ehrhardt, K.; Tallet, C.; Warenghem, M.; Baron, A.; Aradian, A.; Kildemo, M.; Ponsinet, V. Hyperbolic-by-design self-assembled metamaterial based on block copolymers lamellar phases. *Opt. Laser Technol.* **2017**, *88*, 85–95. [CrossRef]
32. Ang, L.Y.L.; Koh, Y.K.; Lee, H.P. Plate-type acoustic metamaterials: Experimental evaluation of a modular large-scale design for low-frequency noise control. *Acoustics* **2019**, *1*, 354–368. [CrossRef]
33. Wang, Q.; Du, H.; Li, F.; Ling, D. Nano-Metamaterial: A State-of-the-Art Material for Magnetic Resonance Imaging. *Small Sci.* **2023**, *3*, 2300015. [CrossRef]
34. Pendry, J.B.; Holden, A.J.; Robbins, D.J.; Stewart, W.J. Magnetism from conductors and enhanced nonlinear phenomena. *IEEE Trans. Microw. Theory Tech.* **1999**, *47*, 2075–2084. [CrossRef]
35. Tzarouchis, D.C.; Koutsoupidou, M.; Sotiriou, I.; Dovelos, K.; Rompolas, D.; Kosmas, P. Electromagnetic metamaterials for biomedical applications: Short review and trends. *EPJ Appl. Metamater.* **2024**, *11*, 7. [CrossRef]
36. Liu, J.; Jennings, S.F.; Tong, W.; Hong, H. Next generation sequencing for profiling expression of miRNAs: Technical progress and applications in drug development. *J. Biomed. Sci. Eng.* **2011**, *4*, 666. [CrossRef] [PubMed]
37. Slobozhanyuk, A.P.; Poddubny, A.N.; Raaijmakers, A.J.; Van Den Berg, C.A.; Kozachenko, A.V.; Dubrovina, I.A.; Melchakova, I.V.; Kivshar, Y.S.; Belov, P.A. Enhancement of magnetic resonance imaging with metasurfaces. *Adv. Mater.* **2016**, *28*, 1832–1838. [CrossRef] [PubMed]
38. Velazquez-Ahumada, M.C.; Freire, M.J.; Marques, R. Metamaterial applicator for microwave hyperthermia. In Proceedings of the 2011 XXXth URSI General Assembly and Scientific Symposium, Istanbul, Turkey, 13–20 August 2011; pp. 1–4.
39. Melik, R.; Unal, E.; Perkgoz, N.K.; Puttlitz, C.; Demir, H.V. Metamaterial-based wireless strain sensors. *Appl. Phys. Lett.* **2009**, *95*, 011106. [CrossRef]
40. Padilla, W.J.; Basov, D.N.; Smith, D.R. Negative refractive index metamaterials. *Mater. Today* **2006**, *9*, 28–35. [CrossRef]
41. Harinarayana, V.; Shin, Y. Two-photon lithography for three-dimensional fabrication in micro/nanoscale regime: A comprehensive review. *Opt. Laser Technol.* **2021**, *142*, 107180. [CrossRef]
42. Leonard, J. Parallel Plate Wave Transmission Structures for Metamaterial Research. *J. Appl. Phys.* **1968**, *39*, 1234–1240.

43. Pendry, J.B.; Smith, D.R. Reversing light with negative refraction. *Phys. Today* **2004**, *57*, 37–43. [CrossRef]
44. Lvov, V.A.; Senatov, F.S.; Veveris, A.A.; Skrybykina, V.A.; Díaz Lantada, A. Auxetic metamaterials for biomedical devices: Current situation, main challenges, and research trends. *Materials* **2022**, *15*, 1439. [CrossRef]
45. Zadpoor, A.A. Design for additive bio-manufacturing: From patient-specific medical devices to rationally designed meta-biomaterials. *Int. J. Mol. Sci.* **2017**, *18*, 1607. [CrossRef]
46. Zhang, K.; Soh, P.J.; Yan, S. Meta-wearable antennas-a review of metamaterial based antennas in wireless body area networks. *Materials* **2021**, *14*, 149. [CrossRef]
47. Salami, P.; Yousefi, L. Wide-band polarisation-independent metasurface-based carpet cloak. *IET Microw. Antennas Propag.* **2020**, *14*, 1983–1989. [CrossRef]
48. Hu, X.; Luo, Y.; Wang, J.; Tang, J.; Gao, Y.; Ren, J.; Yu, H.; Zhang, J.; Ye, D. Multiband omnidirectional invisibility cloak. *Adv. Sci.* **2024**, *11*, 2401295. [CrossRef] [PubMed]
49. Cui, T.J.; Smith, D.R.; Liu, R. *Metamaterials: Theory, Design, and Applications*; Springer: New York, NY, USA, 2010.
50. Holliman, J.E.; Schaef, H.T.; McGrail, B.P.; Miller, Q.R. Review of foundational concepts and emerging directions in metamaterial research: Design, phenomena, and applications. *Mater. Adv.* **2022**, *3*, 8390–8406. [CrossRef]
51. Arjunan, A.; Baroutaji, A.; Robinson, J.; Vance, A.; Arafat, A. Acoustic metamaterials for sound absorption and insulation in buildings. *Build. Environ.* **2024**, *251*, 111250. [CrossRef]
52. Tang, W.; Mei, Z.; Cui, T. Theory, experiment and applications of metamaterials. *Sci. China Phys. Mech. Astron.* **2015**, *58*, 1–11. [CrossRef]
53. Cui, T.J.; Li, L.; Liu, S.; Ma, Q.; Zhang, L.; Wan, X.; Jiang, W.X.; Cheng, Q. Information metamaterial systems. *iScience* **2020**, *23*, 101403. [CrossRef]
54. Liu, Y.; Dong, T.; Qin, X.; Luo, W.; Leng, N.; He, Y.; Yuan, Y.; Bai, M.; Sun, J.; Zhou, J. High-permittivity ceramics enabled highly homogeneous zero-index metamaterials for high-directivity antennas and beyond. *eLight* **2024**, *4*, 4. [CrossRef]
55. Giri, R.; Payal, R. Negative-Index Metamaterials. In *Electromagnetic Metamaterials: Properties and Applications*; Scrivener Publishing LLC: Beverly, MA, USA, 2023; pp. 205–217.
56. Liu, Y.; Wang, Y.; Wang, Q.; Guo, S. A Dual-polarized Broadband Fabry-Perot Antenna Loaded with an Energy-selective Surface. In Proceedings of the 2024 IEEE 7th International Conference on Electronic Information and Communication Technology (ICEICT), Nanjing, China, 15–17 April 2024; pp. 1496–1498.
57. Liu, M.; Lan, X.; Zhang, H.; Xie, P.; Wu, N.; Yuan, H.; Sui, K.; Fan, R.; Liu, C. Iron/epoxy random metamaterials with adjustable epsilon-near-zero and epsilon-negative property. *J. Mater. Sci. Mater. Electron.* **2021**, *32*, 15995–16007. [CrossRef]
58. Gholipur, R.; Bahari, A. Random nanocomposites as metamaterials: Preparation and investigations at microwave region. *Opt. Mater.* **2015**, *50*, 175–183. [CrossRef]
59. Kadic, M.; Bückmann, T.; Schittny, R.; Wegener, M. Metamaterials beyond electromagnetism. *Rep. Prog. Phys.* **2013**, *76*, 126501. [CrossRef]
60. An, H.; Liu, G.; Li, Y.; Song, J.; Zhang, C.; Liu, M. Inhomogeneous electromagnetic metamaterial design method for improving efficiency and range of wireless power transfer. *IET Microw. Antennas Propag.* **2019**, *13*, 2110–2118. [CrossRef]
61. Cui, T.J.; Qi, M.Q.; Wan, X.; Zhao, J.; Cheng, Q. Coding metamaterials, digital metamaterials and programmable metamaterials. *Light Sci. Appl.* **2014**, *3*, e218. [CrossRef]
62. Choi, J.R. Analysis of light-wave nonstaticity in the coherent state. *Sci. Rep.* **2021**, *11*, 23974. [CrossRef]
63. McPhedran, R.C.; Shadrivov, I.V.; Kuhlmey, B.T.; Kivshar, Y.S. Metamaterials and metaoptics. *NPG Asia Mater.* **2011**, *3*, 100–108. [CrossRef]
64. Cai, W.; Shalaev, V. *Optical Metamaterials*; Fundamentals and Applications; Springer: New York, NY, USA, 2011.
65. Zhang, D.; Ren, J.; Zhou, T.; Li, B. Dark state, zero-index and topology in phononic metamaterials with negative mass and negative coupling. *New J. Phys.* **2019**, *21*, 093033. [CrossRef]
66. Xiao, S.; Drachev, V.P.; Kildishev, A.V.; Ni, X.; Chettiar, U.K.; Yuan, H.-K.; Shalaev, V.M. Loss-free and active optical negative-index metamaterials. *Nature* **2010**, *466*, 735–738. [CrossRef] [PubMed]
67. Liu, Y.; Zhang, X. Metamaterials: A new frontier of science and technology. *Chem. Soc. Rev.* **2011**, *40*, 2494–2507. [CrossRef]
68. Shalaev, V.M. Optical negative-index metamaterials. *Nat. Photonics* **2007**, *1*, 41–48. [CrossRef]
69. Zhang, B. Electrodynamics of transformation-based invisibility cloaking. *Light Sci. Appl.* **2012**, *1*, e32. [CrossRef]
70. Zhang, L.; Koschny, T.; Soukoulis, C.M. Creating double negative index materials using the Babinet principle with one metasurface. *Phys. Rev. B—Condens. Matter Mater. Phys.* **2013**, *87*, 045101. [CrossRef]
71. Wu, X.; Su, Y.; Shi, J. Perspective of additive manufacturing for metamaterials development. *Smart Mater. Struct.* **2019**, *28*, 093001. [CrossRef]
72. Wang, L.; Karaaslan, M. *Advanced Metamaterials for Engineers*; IOP Publishing: Bristol, UK, 2023.
73. Sidhwa, H.H. Application of transformation optics for the purpose of cloaking. *Phys. Astron. Int. J.* **2018**, *2*, 111–114.

74. Zhu, T.; Zhou, Y.; Lou, Y.; Ye, H.; Qiu, M.; Ruan, Z.; Fan, S. Plasmonic computing of spatial differentiation. *Nat. Commun.* **2017**, *8*, 15391. [CrossRef]
75. Kanamori, Y.; Hokari, R.; Hane, K. MEMS for plasmon control of optical metamaterials. *IEEE J. Sel. Top. Quantum Electron.* **2015**, *21*, 137–146. [CrossRef]
76. Takaloo, S.; Xu, A.H.; Zaidan, L.; Irannejad, M.; Yavuz, M. Towards Point-of-Care Single Biomolecule Detection Using Next Generation Portable Nanoplasmonic Biosensors: A Review. *Biosensors* **2024**, *14*, 593. [CrossRef]
77. Xu, Y.; Sun, M.; Wu, H.; Song, Y.; Wang, Q. Plasmonic biosensor based on Ag-TiO 2-ZnO gratings for cancer detection in the optical communication band. *IEEE Sens. J.* **2023**, *23*, 20959–20967. [CrossRef]
78. Eissa, M.E. Nanosensors for early detection and diagnosis of cancer: A review of recent advances. *J. Cancer Res. Rev.* **2024**, *1*, 1–13. [CrossRef]
79. Fatema, M. Advances in Nanomaterial-Based Sensors for Early Disease Detection. *Int. J. Adv. Res. Sci. Commun. Technol.* **2024**, *10*, 653–658. [CrossRef]
80. Cummer, S.A.; Christensen, J.; Alù, A. Controlling sound with acoustic metamaterials. *Nat. Rev. Mater.* **2016**, *1*, 16001. [CrossRef]
81. Bertoldi, K.; Vitelli, V.; Christensen, J.; Van Hecke, M. Flexible mechanical metamaterials. *Nat. Rev. Mater.* **2017**, *2*, 17066. [CrossRef]
82. Lakes, R. Foam structures with a negative Poisson's ratio. *Science* **1987**, *235*, 1038–1040. [CrossRef] [PubMed]
83. Dede, E.M.; Zhou, F.; Schmalenberg, P.; Nomura, T. Thermal metamaterials for heat flow control in electronics. *J. Electron. Packag.* **2018**, *140*, 010904. [CrossRef]
84. Zadpoor, A.A.; Mirzaali, M.J.; Valdevit, L.; Hopkins, J.B. Design, material, function, and fabrication of metamaterials. *APL Mater.* **2023**, *11*, 020401. [CrossRef]
85. Zheng, X.; Lee, H.; Weisgraber, T.H.; Shusteff, M.; DeOtte, J.; Duoss, E.B.; Kuntz, J.D.; Biener, M.M.; Ge, Q.; Jackson, J.A. Ultralight, ultrastiff mechanical metamaterials. *Science* **2014**, *344*, 1373–1377. [CrossRef]
86. Chanda, D.; Shigeta, K.; Gupta, S.; Cain, T.; Carlson, A.; Mihi, A.; Baca, A.J.; Bogart, G.R.; Braun, P.; Rogers, J.A. Large-area flexible 3D optical negative index metamaterial formed by nanotransfer printing. *Nat. Nanotechnol.* **2011**, *6*, 402–407. [CrossRef]
87. Nandja, S. Light control with negative index metamaterial and phase change material at optical wavelength. *Optik* **2022**, *267*, 169742. [CrossRef]
88. Yu, M.; Wang, Y.; Zhong, W.; Guo, R.; Zhou, X. Optical properties of strongly anisotropic metamaterials. *Appl. Phys. A* **2012**, *108*, 65–73. [CrossRef]
89. Jia, X.; Yan, M.; Hong, M. Sound energy enhancement via impedance-matched anisotropic metamaterial. *Mater. Des.* **2021**, *197*, 109254. [CrossRef]
90. Chang, Y.; Wei, J.; Lee, C. Metamaterials–from fundamentals and MEMS tuning mechanisms to applications. *Nanophotonics* **2020**, *9*, 3049–3070. [CrossRef]
91. Grimberg, R. Electromagnetic metamaterials. *Mater. Sci. Eng. B* **2013**, *178*, 1285–1295. [CrossRef]
92. RoyChoudhury, S.; Rawat, V.; Jalal, A.H.; Kale, S.; Bhansali, S. Recent advances in metamaterial split-ring-resonator circuits as biosensors and therapeutic agents. *Biosens. Bioelectron.* **2016**, *86*, 595–608. [CrossRef] [PubMed]
93. Xia, Y.; Erturk, A.; Ruzzene, M. Topological edge states in quasiperiodic locally resonant metastructures. *Phys. Rev. Appl.* **2020**, *13*, 014023. [CrossRef]
94. Contreras, N.; Zhang, X.; Hao, H.; Hernández, F. Application of elastic metamaterials/meta-structures in civil engineering: A review. *Compos. Struct.* **2024**, *327*, 117663. [CrossRef]
95. Besteiro, L.V.; Yu, P.; Wang, Z.; Holleitner, A.W.; Hartland, G.V.; Wiederrecht, G.P.; Govorov, A.O. The fast and the furious: Ultrafast hot electrons in plasmonic metastructures. Size and structure matter. *Nano Today* **2019**, *27*, 120–145. [CrossRef]
96. Khanikaev, A.B.; Wu, C.; Shvets, G. Fano-resonant metamaterials and their applications. *Nanophotonics* **2013**, *2*, 247–264. [CrossRef]
97. Feng, Y.; Liang, M.; Zhao, X.; You, R. Fabrication and modulation of flexible electromagnetic metamaterials. *Microsyst. Nanoeng.* **2025**, *11*, 14. [CrossRef]
98. Li, X.; Pu, M.; Ma, X.; Guo, Y.; Gao, P.; Luo, X. Dispersion engineering in metamaterials and metasurfaces. *J. Phys. D Appl. Phys.* **2018**, *51*, 054002. [CrossRef]
99. Teixeira, F.; Sarris, C.; Zhang, Y.; Na, D.-Y.; Berenger, J.-P.; Su, Y.; Okoniewski, M.; Chew, W.; Backman, V.; Simpson, J.J. Finite-difference time-domain methods. *Nat. Rev. Methods Primers* **2023**, *3*, 75. [CrossRef]
100. Yang, P.; Liou, K. Finite-difference time domain method for light scattering by small ice crystals in three-dimensional space. *J. Opt. Soc. Am. A* **1996**, *13*, 2072–2085. [CrossRef]
101. Taravati, S. Generalized FDTD Numerical Modeling of Space-Time-Varying Media. In Proceedings of the 2024 54th European Microwave Conference (EuMC), Paris, France, 24–26 September 2024; pp. 920–923.
102. Huang, Y.; Li, J.; Yang, W. Modeling backward wave propagation in metamaterials by the finite element time-domain method. *SIAM J. Sci. Comput.* **2013**, *35*, B248–B274. [CrossRef]

103. David Müzel, S.; Bonhin, E.P.; Guimarães, N.M.; Guidi, E.S. Application of the finite element method in the analysis of composite materials: A review. *Polymers* **2020**, *12*, 818. [CrossRef] [PubMed]
104. Wang, K.X.; Zhou, E.L.; Wei, B.L.; Wu, Y.; Wang, G. An efficient and accurate numerical method for the heat conduction problems of thermal metamaterials based on edge-based smoothed finite element method. *Eng. Anal. Bound. Elem.* **2022**, *134*, 282–297. [CrossRef]
105. Weismann, M.; Gallagher, D.F.; Panoiu, N.C. Accurate near-field calculation in the rigorous coupled-wave analysis method. *J. Opt.* **2015**, *17*, 125612. [CrossRef]
106. Tihon, D.; Sozio, V.; Ozdemir, N.A.; Albani, M.; Craeye, C. Numerically stable eigenmode extraction in 3-D periodic metamaterials. *IEEE Trans. Antennas Propag.* **2016**, *64*, 3068–3079. [CrossRef]
107. Barbarić, D.; Bosiljevac, M.; Šipuš, Z. Analysis of curved metasurfaces based on method of moments. In Proceedings of the 2020 14th European Conference on Antennas and Propagation (EuCAP), Copenhagen, Denmark, 15–20 March 2020; pp. 1–5.
108. Bashir, I.; Carley, M. Development of 3D boundary element method for the simulation of acoustic metamaterials/metasurfaces in mean flow for aerospace applications. *Int. J. Aeroacoustics* **2020**, *19*, 324–346. [CrossRef]
109. Hamza, M.N.; Koziel, S.; Pietrenko-Dabrowska, A. Design and experimental validation of a metamaterial-based sensor for microwave imaging in breast, lung, and brain cancer detection. *Sci. Rep.* **2024**, *14*, 16177. [CrossRef] [PubMed]
110. Chronopoulos, D.; Meng, H.; Elmadih, W.; Fabro, A.; Maskery, I. Rainbow metamaterials for broadband multi-frequency vibration attenuation: Numerical analysis and experimental validation. *J. Sound Vib.* **2020**, *465*, 115005.
111. Hermann, S.; Billon, K.; Parlak, A.-M.; Orlowsky, J.; Collet, M.; Madeo, A. Design and experimental validation of a finite-size labyrinthine metamaterial for vibro-acoustics: Enabling upscaling towards large-scale structures. *Philos. Trans. A* **2024**, *382*, 20230367. [CrossRef]
112. Fischer, S.C.; Hillen, L.; Eberl, C. Mechanical metamaterials on the way from laboratory scale to industrial applications: Challenges for characterization and scalability. *Materials* **2020**, *13*, 3605. [CrossRef]
113. Zhu, S.; Cao, Y.; Fu, Y.; Gao, L.; Li, X.; Chen, H.; Xu, Y. 3D broadband waveguide cloak and light squeezing in terahertz regime. *Opt. Lett.* **2020**, *45*, 652–655. [CrossRef]
114. Zheng, X.; Zhang, X.; Chen, T.T.; Watanabe, I. Deep learning in mechanical metamaterials: From prediction and generation to inverse design. *Adv. Mater.* **2023**, *35*, 2302530. [CrossRef] [PubMed]
115. Chen, W.; Sun, R.; Lee, D.; Portela, C.M.; Chen, W. Generative Inverse Design of Metamaterials with Functional Responses by Interpretable Learning. *Adv. Intell. Syst.* **2024**, 2400611. [CrossRef]
116. Ha, C.S.; Yao, D.; Xu, Z.; Liu, C.; Liu, H.; Elkins, D.; Kile, M.; Deshpande, V.; Kong, Z.; Bauchy, M. Rapid inverse design of metamaterials based on prescribed mechanical behavior through machine learning. *Nat. Commun.* **2023**, *14*, 5765. [CrossRef] [PubMed]
117. Nezaratizadeh, A.; Hashemi, S.M.; Bod, M. Deep Learning for Electromagnetic Metamaterial Inverse Design. In Proceedings of the 2024 11th International Symposium on Telecommunications (IST), Tehran, Iran, 9–10 October 2024; pp. 279–282.
118. Xiao, X.; Chen, J.; Wang, K.; Yu, Y.; Wei, K. Multimaterial additively manufactured metamaterials functionalized with customizable thermal expansion in multiple directions. *ACS Appl. Mater. Interfaces* **2023**, *15*, 47434–47446. [CrossRef] [PubMed]
119. Askari, M.; Hutchins, D.A.; Thomas, P.J.; Astolfi, L.; Watson, R.L.; Abdi, M.; Ricci, M.; Laureti, S.; Nie, L.; Freear, S. Additive manufacturing of metamaterials: A review. *Addit. Manuf.* **2020**, *36*, 101562. [CrossRef]
120. El Helou, C.; Buskohl, P.R.; Tabor, C.E.; Harne, R.L. Digital logic gates in soft, conductive mechanical metamaterials. *Nat. Commun.* **2021**, *12*, 1633. [CrossRef]
121. Jiao, P.; Mueller, J.; Raney, J.R.; Zheng, X.; Alavi, A.H. Mechanical metamaterials and beyond. *Nat. Commun.* **2023**, *14*, 6004. [CrossRef]
122. Bowen, P.T.; Driscoll, T.; Kundtz, N.B.; Smith, D.R. Using a discrete dipole approximation to predict complete scattering of complicated metamaterials. *New J. Phys.* **2012**, *14*, 033038. [CrossRef]
123. Jiao, P.; Alavi, A.H. Artificial intelligence-enabled smart mechanical metamaterials: Advent and future trends. *Int. Mater. Rev.* **2021**, *66*, 365–393. [CrossRef]
124. Fu, K.; Zhao, Z.; Jin, L. Programmable granular metamaterials for reusable energy absorption. *Adv. Funct. Mater.* **2019**, *29*, 1901258. [CrossRef]
125. Dora, E.A. *Advances in Metamaterials*; N Y Research Press: Forest Hills, NY, USA, 2017; Chapter 1.
126. Li, B.; Zhao, Z.; Song, L.; Hu, R.; Fan, L.; Xie, J.; Du, X.; Wang, M. Numerical simulation and electromagnetic parameter retrieve: Performance evaluation of metamaterials under TE and TM polarization conditions. *Model. Simul. Mater. Sci. Eng.* **2025**, *33*, 015015. [CrossRef]
127. Tong, X.C. *Functional Metamaterials and Metadevices*; Springer: Cham, Switzerland, 2018; Chapter 2.
128. Jeppesen, C.; Mortensen, N.A.; Kristensen, A. Capacitance tuning of nanoscale split-ring resonators. *Appl. Phys. Lett.* **2009**, *95*, 193108. [CrossRef]

129. Sydoruk, O.; Tatartschuk, E.; Shamonina, E.; Solymar, L. Analytical formulation for the resonant frequency of split rings. *J. Appl. Phys.* **2009**, *105*, 014903. [CrossRef]
130. Menzel, C.; Helgert, C.; Rockstuhl, C.; Kley, E.B.; Tünnermann, A.; Pertsch, T.; Lederer, F. Asymmetric Transmission of Linearly Polarized Light at Optical Metamaterials. *Phys. Rev. Lett.* **2010**, *104*, 253902. [CrossRef]
131. Huang, C.; Feng, Y.; Zhao, J.; Wang, Z.; Jiang, T. Asymmetric electromagnetic wave transmission of linear polarization via polarization conversion through chiral metamaterial structures. *Phys. Rev. B* **2012**, *85*, 195131. [CrossRef]
132. Picozzi, A.; Millot, G.; Wabnitz, S. Nonlinear virtues of multimode fibre. *Nat. Photonics* **2015**, *9*, 289–291. [CrossRef]
133. Maraghechi, S.; Hoefnagels, J.P.M.; Peerlings, R.H.J.; Rokoš, O.; Geers, M.G.D. Experimental full-field analysis of size effects in miniaturized cellular elastomeric metamaterials. *Mater. Des.* **2020**, *193*, 108684. [CrossRef]
134. Rao, K.N.; Laxmareddy, D.; Monika, S.; Sailaja, M. Design of Unit Cells-Based Metamaterial Antenna. *J. Phys. Conf. Ser.* **2024**, *2837*, 012013.
135. Abdulkarim, Y.I.; Deng, L.; Luo, H.; Huang, S.; Karaaslan, M.; Altintas, O.; Bakir, M.; Muhammadsharif, F.F.; Awl, H.N.; Sabah, C.; et al. Design and study of a metamaterial based sensor for the application of liquid chemicals detection. *J. Mater. Res. Technol.* **2020**, *9*, 10291–10304. [CrossRef]
136. Khansanami, M.; Younesian, D. A Novel Unit Cell for Low-Frequency Vibration Suppression Through Meta-Plates: Modeling, Optimization and Testing. *Int. J. Appl. Mech.* **2022**, *15*, 2250079. [CrossRef]
137. Gozhenko, V.V.; Amert, A.K.; Whites, K.W. Homogenization of periodic metamaterials by field averaging over unit cell boundaries: Use and limitations. *New J. Phys.* **2013**, *15*, 043030. [CrossRef]
138. Pitilakis, A.; Tsilipakos, O.; Liu, F.; Kossifos, K.M.; Tasolamprou, A.C.; Kwon, D.H.; Mirmoosa, M.S.; Manessis, D.; Kantartzis, N.V.; Liaskos, C.; et al. A Multi-Functional Reconfigurable Metasurface: Electromagnetic Design Accounting for Fabrication Aspects. *IEEE Trans. Antennas Propag.* **2021**, *69*, 1440–1454. [CrossRef]
139. Silveirinha, M.t.a.G. Metamaterial homogenization approach with application to the characterization of microstructured composites with negative parameters. *Phys. Rev. B* **2007**, *75*, 115104. [CrossRef]
140. Wang, C.; Vangelatos, Z.; Grigoropoulos, C.P.; Ma, Z. Micro-engineered architected metamaterials for cell and tissue engineering. *Mater. Today Adv.* **2022**, *13*, 100206. [CrossRef]
141. Rajasri, S.; Rani, R.B. Design and Performance Analysis of Metamaterial-Inspired Decagon-Shaped Antenna for Vehicular Communications. *Prog. Electromagn. Res. Lett.* **2022**, *105*, 139–147. [CrossRef]
142. Gozhenko, V.V. Size Effects in Periodic Metamaterials. *arXiv* **2023**, arXiv:2301.03518.
143. Rajasri, S.; Rani, B.R. *Analysis of Unit Cell with and Without Splits for Understanding Metamaterial Property*; Springer: Singapore, 2022. [CrossRef]
144. Turkmen, O.; Ekmekci, E.; Turhan-Sayan, G. Nested U-ring resonators: A novel multi-band metamaterial design in microwave region. *IET Microw. Antennas Propag.* **2012**, *6*, 1102–1108. [CrossRef]
145. Savo, S.; Shrekenhamer, D.; Padilla, W.J. Liquid crystal metamaterial absorber spatial light modulator for THz applications. *Adv. Opt. Mater.* **2014**, *2*, 275–279. [CrossRef]
146. Pitchappa, P.; Ho, C.P.; Dhakar, L.; Lee, C. Microelectromechanically reconfigurable interpixelated metamaterial for independent tuning of multiple resonances at terahertz spectral region. *Optica* **2015**, *2*, 571. [CrossRef]
147. Pitchappa, P.; Ho, C.P.; Cong, L.; Singh, R.; Singh, N.; Lee, C. Reconfigurable Digital Metamaterial for Dynamic Switching of Terahertz Anisotropy. *Adv. Opt. Mater.* **2016**, *4*, 391–398. [CrossRef]
148. Vafapour, Z.; Troy, W.; Rashidi, A. Colon Cancer Detection by Designing and Analytical Evaluation of a Water-Based THz Metamaterial Perfect Absorber. *IEEE Sens. J.* **2021**, *21*, 19307–19313. [CrossRef]
149. Jia, D.; He, Y.; Ding, N.; Zhou, J.; Du, B.; Zhang, W. Beam-Steering Flat Lens Antenna Based on Multilayer Gradient Index Metamaterials. *IEEE Antennas Wirel. Propag. Lett.* **2018**, *17*, 1510–1514. [CrossRef]
150. Xu, J.; Yang, R.; Fan, Y.; Fu, Q.; Zhang, F. A Review of Tunable Electromagnetic Metamaterials With Anisotropic Liquid Crystals. *Front. Phys.* **2021**, *9*, 633104. [CrossRef]
151. Ben-Yelun, I.; Saucedo-Mora, L.; Sanz, M.Á.; Benítez, J.M.; Montans, F.J. Topology optimization approach for functionally graded metamaterial components based on homogenization of mechanical variables. *Comput. Struct.* **2023**, *289*, 107151. [CrossRef]
152. Wu, J.; Sigmund, O.; Groen, J.P. Topology optimization of multi-scale structures: A review. *Struct. Multidiscip. Optim.* **2021**, *63*, 1455–1480. [CrossRef]
153. Sigmund, O.; Torquato, S. Design of smart composite materials using topology optimization. *Smart Mater. Struct.* **1999**, *8*, 365–379. [CrossRef]
154. Zheng, R.; Yi, B.; Tao, Y.; Peng, X. Topology optimization of extreme mechanical metamaterials considering the anisotropy of additive manufactured parts. *Smart Mater. Struct.* **2024**, *33*, 115024. [CrossRef]
155. Sigmund, O. Systematic Design of Metamaterials by Topology Optimization. In *IUTAM Symposium on Modelling Nanomaterials and Nanosystems*; IUTAM Bookseries; Springer: Dordrecht, The Netherlands, 2009; Volume 13.

156. Sha, W.; Xiao, M.; Wang, Y.; Huang, M.; Li, Q.; Gao, L. Topology optimization methods for thermal metamaterials: A review. *Int. J. Heat Mass Transf.* **2024**, *227*, 125588. [CrossRef]
157. Wu, W.; Wang, Y.; Gao, Z.; Liu, P. Topology Optimization of Metamaterial Microstructures for Negative Poisson's Ratio under Large Deformation Using a Gradient-Free Method. *Comput. Model. Eng. Sci.* **2024**, *139*, 2001–2026. [CrossRef]
158. Murai, N.; Noguchi, Y.; Matsushima, K.; Yamada, T. Multiscale topology optimization of electromagnetic metamaterials using a high-contrast homogenization method. *Comput. Methods Appl. Mech. Eng.* **2023**, *403*, 115728. [CrossRef]
159. Ishikawa, A.; Tanaka, T. Three-dimensional plasmonic metamaterials and their fabrication techniques. *IEEE J. Sel. Top. Quantum Electron.* **2013**, *19*, 4700312. [CrossRef]
160. Suárez, L.; del Mar Espinosa, M. Assessment on the use of additive manufacturing technologies for acoustic applications. *Int. J. Adv. Manuf. Technol.* **2020**, *109*, 2691–2705. [CrossRef]
161. Bruns, T.E. A reevaluation of the SIMP method with filtering and an alternative formulation for solid-void topology optimization. *Struct. Multidiscip. Optim.* **2005**, *30*, 428–436. [CrossRef]
162. Sanfui, S.; Sharma, D. Soft- and Hard-Kill Hybrid GPU-based Bi-Directional Evolutionary Structural Optimization. *J. Comput. Inf. Sci. Eng.* **2024**, *24*, 041008. [CrossRef]
163. Ghabraie, K. An improved soft-kill BESO algorithm for optimal distribution of single or multiple material phases. *Struct. Multidiscip. Optim.* **2015**, *52*, 773–790. [CrossRef]
164. Liu, S.; Li, Q.; Hu, J.; Chen, W.; Zhang, Y.; Luo, Y.; Wang, Q. A Survey of Topology Optimization Methods Considering Manufacturable Structural Feature Constraints for Additive Manufacturing Structures. *Addit. Manuf. Front.* **2024**, *3*, 200143. [CrossRef]
165. Kumar, P.K.A.V.; Li, P.; Reinoso, J.; He, Q.C.; Yvonnet, J.; Paggi, M. SIMP Phase-field topology optimization framework to maximize fracture resistance in FGMs. *Compos. Struct.* **2024**, *329*, 117750. [CrossRef]
166. Xue, H.; Yu, H.; Zhang, X.; Quan, Q. A novel method for structural lightweight design with topology optimization. *Energies* **2021**, *14*, 4367. [CrossRef]
167. Luo, Y.; Sigmund, O.; Li, Q.; Liu, S. Topology optimization of structures with infill-supported enclosed voids for additive manufacturing. *Addit. Manuf.* **2022**, *55*, 102795. [CrossRef]
168. Liu, Y.; Wang, Y.; Ren, H.; Meng, Z.; Chen, X.; Li, Z.; Wang, L.; Chen, W.; Wang, Y.; Du, J. Ultrastiff metamaterials generated through a multilayer strategy and topology optimization. *Nat. Commun.* **2024**, *15*, 47089. [CrossRef] [PubMed]
169. Akhmetshin, L.; Iokhim, K.; Kazantseva, E.; Smolin, I. Influence of Topological Defects on the Mechanical Response of Unit Cells of the Tetrachiral Mechanical Metamaterial. *Designs* **2023**, *7*, 129. [CrossRef]
170. Zeng, Q.; Duan, S.; Zhao, Z.; Wang, P.; Lei, H. Inverse Design of Energy-Absorbing Metamaterials by Topology Optimization. *Adv. Sci.* **2023**, *10*, 2204977. [CrossRef]
171. Stepanova, M.; Dew, S. (Eds.) *Nanofabrication: Techniques and Principles*; Springer: Vienna, Austria, 2012.
172. Feng, F.; He, C.; Cui, Z.; Ying, T.; Cai, J.; Tao, M. Topology Optimization of Multi-Material Underwater Broadband Sound Absorption Metamaterial Based on Genetic Algorithm. Available online: https://ouci.dntb.gov.ua/en/works/7PP83wQ7/ (accessed on 5 April 2025).
173. Zheng, R.; Yi, B.; Liu, W.; Liu, L.; Peng, X.; Tao, Y. Topology optimization of self-supporting metamaterials for additive manufacturing: A novel framework and validation. *Smart Mater. Struct.* **2025**, *34*, 015056. [CrossRef]
174. Viswanath, A.; Abueidda, D.W.; Modrek, M.; Al-Rub, R.K.A.; Koric, S.; Khan, K.A. Designing a TPMS metamaterial via deep learning and topology optimization. *Front. Mech. Eng.* **2024**, *10*, 1417606. [CrossRef]
175. Qin, H.; Yang, D.; Ren, C. Design method of lightweight metamaterials with arbitrary Poisson's ratio. *Materials* **2018**, *11*, 1574. [CrossRef]
176. Sha, W.; Xiao, M.; Zhang, J.; Ren, X.; Zhu, Z.; Zhang, Y.; Xu, G.; Li, H.; Liu, X.; Chen, X.; et al. Robustly printable freeform thermal metamaterials. *Nat. Commun.* **2021**, *12*, 27543. [CrossRef]
177. Gustavo Méndez, C.; Podestá, J.M.; Lloberas-Valls, O.; Toro, S.; Huespe, A.E.; Oliver, J. Computational material design for acoustic cloaking. *Int. J. Numer. Methods Eng.* **2017**, *112*, 1353–1380. [CrossRef]
178. Stankiewicz, G.; Dev, C.; Weichelt, M.; Fey, T.; Steinmann, P. Towards advanced piezoelectric metamaterial design via combined topology and shape optimization. *Res. Sq.* **2023**, *67*, 26. [CrossRef]
179. Han, Z.; Xiao, X.; Chen, J.; Wei, K.; Wang, Z.; Yang, X.; Fang, D. Bifunctional Metamaterials Incorporating Unusual Geminations of Poisson's Ratio and Coefficient of Thermal Expansion. *ACS Appl. Mater. Interfaces* **2022**, *14*, 50068–50078. [CrossRef] [PubMed]
180. Chen, Z.; Lin, Y.T.; Salehi, H.; Che, Z.; Zhu, Y.; Ding, J.; Sheng, B.; Zhu, R.; Jiao, P. Advanced Fabrication of Mechanical Metamaterials Based on Micro/Nanoscale Technology. *Adv. Eng. Mater.* **2023**, *22*, 202300750.
181. Baracu, A.M.; Avram, M.A.; Breazu, C.; Bunea, M.C.; Socol, M.; Stanculescu, A.; Matei, E.; Thrane, P.C.V.; Dirdal, C.A.; Dinescu, A.; et al. Silicon metalens fabrication from electron beam to uvnanoimprint lithography. *Nanomaterials* **2021**, *11*, 2329. [CrossRef] [PubMed]

182. Leontiev, A.P.; Sotnichuk, S.V.; Klimenko, A.A.; Malysheva, I.V.; Kolmychek, I.A.; Mumlyakov, A.M.; Tsiniaikin, I.I.; Murzina, T.V.; Napolskii, K.S. Ion beam etching of anodic aluminium oxide barrier layer for Au nanorod-based hyperbolic metamaterials. *J. Mater. Chem. C* **2024**, *12*, 9274–9283. [CrossRef]
183. Park, S.J.; Cunningham, J. Effect of substrate etching on terahertz metamaterial resonances and its liquid sensing applications. *Sensors* **2020**, *20*, 3133. [CrossRef] [PubMed]
184. Chen, Y. Nanofabrication by electron beam lithography and its applications: A review. *Microelectron. Eng.* **2015**, *135*, 57–72. [CrossRef]
185. Tan, Y.S.; Wang, H.; Wang, H.; Pan, C.; Yang, J.K.W. High-throughput fabrication of large-scale metasurfaces using electron-beam lithography with SU-8 gratings for multilevel security printing. *Photonics Res.* **2023**, *11*, B103. [CrossRef]
186. Huff, M. Recent advances in reactive ion etching and applications of high-aspect-ratio microfabrication. *Micromachines* **2021**, *12*, 991. [CrossRef]
187. Baron, A.; Aradian, A.; Ponsinet, V.; Barois, P. Self-assembled optical metamaterials. *Opt. Laser Technol.* **2016**, *82*, 94–100. [CrossRef]
188. Walia, S.; Shah, C.M.; Gutruf, P.; Nili, H.; Chowdhury, D.R.; Withayachumnankul, W.; Bhaskaran, M.; Sriram, S. Flexible metasurfaces and metamaterials: A review of materials and fabrication processes at micro- and nano-scales. *Appl. Phys. Rev.* **2015**, *2*, 011303. [CrossRef]
189. Dogan, E.; Bhusal, A.; Cecen, B.; Miri, A.K. 3D Printing metamaterials towards tissue engineering. *Appl. Mater. Today* **2020**, *20*, 100752. [CrossRef]
190. Loh, L.Y.W.; Gupta, U.; Wang, Y.; Foo, C.C.; Zhu, J.; Lu, W.F. 3D Printed Metamaterial Capacitive Sensing Array for Universal Jamming Gripper and Human Joint Wearables. *Adv. Eng. Mater.* **2021**, *23*, 2001082. [CrossRef]
191. Zhao, X. Bottom-up fabrication methods of optical metamaterials. *J. Mater. Chem.* **2012**, *22*, 9439–9449. [CrossRef]
192. Peng, J.; Wang, S.; Liang, B.; Wen, Q.; Sun, C.; Li, K.; Zhang, X.; Zhang, Y. Review of micro and nano scale 3D printing of electromagnetic metamaterial absorbers: Mechanism, fabrication, and functionality. *Virtual Phys. Prototyp.* **2024**, *19*, e2378937. [CrossRef]
193. Jin, H.; Espinosa, H.D. Mechanical Metamaterials Fabricated From Self-Assembly: A Perspective. *J. Appl. Mech. Trans. ASME* **2024**, *91*, 1–25. [CrossRef]
194. Wang, K.; Park, S.H.; Zhu, J.; Kim, J.K.; Zhang, L.; Yi, G.R. Self-Assembled Colloidal Nanopatterns toward Unnatural Optical Meta-Materials. *Adv. Funct. Mater.* **2021**, *31*, 2008246. [CrossRef]
195. Mühlig, S.; Cunningham, A.; Dintinger, J.; Scharf, T.; Bürgi, T.; Lederer, F.; Rockstuhl, C. Self-assembled plasmonic metamaterials. *Nanophotonics* **2013**, *2*, 211–240. [CrossRef]
196. Wohlwend, J.; Haberfehlner, G.; Galinski, H. Strong Coupling in Two-Phase Metamaterials Fabricated by Sequential Self-Assembly. *Adv. Opt. Mater.* **2023**, *11*, 2300568. [CrossRef]
197. Turek, V.A.; Francescato, Y.; Cadinu, P.; Crick, C.R.; Elliott, L.; Chen, Y.; Urland, V.; Ivanov, A.P.; Velleman, L.; Hong, M.; et al. Self-Assembled Spherical Supercluster Metamaterials from Nanoscale Building Blocks. *ACS Photonics* **2016**, *3*, 35–42. [CrossRef]
198. Yang, W.; Lu, J.; Zhuang, W.; Qi, J.; Wang, C.; Wang, H.; Su, G.; Xiong, K.; Mao, Y.; Gong, X.; et al. PtS2 metamaterials: Fabrication and physical properties. *Appl. Surf. Sci.* **2023**, *614*, 156277. [CrossRef]
199. Sabzi, M.; Anijdan, S.H.M.; Shamsodin, M.; Farzam, M.; Hojjati-Najafabadi, A.; Feng, P.; Park, N.; Lee, U. A Review on Sustainable Manufacturing of Ceramic-Based Thin Films by Chemical Vapor Deposition (CVD): Reactions Kinetics and the Deposition Mechanisms. *Coatings* **2023**, *13*, 188. [CrossRef]
200. Koppes, A.N.; Kamath, M.; Pfluger, C.A.; Burkey, D.D.; Dokmeci, M.; Wang, L.; Carrier, R.L. Complex, multi-scale small intestinal topography replicated in cellular growth substrates fabricated via chemical vapor deposition of Parylene C. *Biofabrication* **2016**, *8*, 035011. [CrossRef]
201. Hosseinzadeh, H.R.S. Metamaterials in Medicine: A New Era for Future Orthopedics. *Orthop. Res. Online J.* **2018**, *2*, 181–183. [CrossRef]
202. Beliaev, L.Y.; Shkondin, E.; Lavrinenko, A.V.; Takayama, O. Optical, structural and composition properties of silicon nitride films deposited by reactive radio-frequency sputtering, low pressure and plasma-enhanced chemical vapor deposition. *Thin Solid Film.* **2022**, *763*, 139568. [CrossRef]
203. Zhang, K.P.; Liao, Y.F.; Qiu, B.; Zheng, Y.K.; Yu, L.K.; He, G.H.; Chen, Q.N.; Sun, D.H. 3D Printed Embedded Metamaterials. *Small* **2021**, *17*, 2103262. [CrossRef]
204. Yuan, S.; Chua, C.K.; Zhou, K. 3D-Printed Mechanical Metamaterials with High Energy Absorption. *Adv. Mater. Technol.* **2019**, *4*, 1800419. [CrossRef]
205. Xie, Y.; Ye, S.; Reyes, C.; Sithikong, P.; Popa, B.I.; Wiley, B.J.; Cummer, S.A. Microwave metamaterials made by fused deposition 3D printing of a highly conductive copper-based filament. *Appl. Phys. Lett.* **2017**, *110*, 181903. [CrossRef]
206. Kennedy, J.; Flanagan, L.; Dowling, L.; Bennett, G.J.; Rice, H.; Trimble, D. The influence of additive manufacturing processes on the performance of a periodic acoustic metamaterial. *Int. J. Polym. Sci.* **2019**, *2019*, 29143. [CrossRef]

207. Yang, D.; Mei, H.; Yao, L.; Yang, W.; Yao, Y.; Cheng, L.; Zhang, L.; Dassios, K.G. *3D/4D Printed Tunable Electrical Metamaterials with More Sophisticated Structures*; Royal Society of Chemistry: London, UK, 2021.
208. Ma, W.W.S.; Yang, H.; Zhao, Y.; Li, X.; Ding, J.; Qu, S.; Liu, Q.; Hu, Z.; Li, R.; Tao, Q.; et al. Multi-Physical Lattice Metamaterials Enabled by Additive Manufacturing: Design Principles, Interaction Mechanisms, and Multifunctional Applications. *Adv. Sci.* **2025**, *12*, 2405835. [CrossRef] [PubMed]
209. Zhou, Y.; Qidong, Y.; Wei, K. Additively manufactured multi-functional metamaterials: Low coefficient of thermal expansion and programmable Poisson's ratio. *Virtual Phys. Prototyp.* **2024**, *19*, e2303714. [CrossRef]
210. Xu, T.; Wang, C.; Luo, X. Interference photolithography with metamaterials. In Proceedings of the 2008 IEEE Photonics-Global@Singapore, Singapore, 8–11 December 2008.
211. Lu, J.; Zhang, X.; Su, G.; Yang, W.; Han, K.; Yu, X.; Wan, Y.; Wang, X.; Yang, P. Large-area uniform few-layer PtS2: Synthesis, structure and physical properties. *Mater. Today Phys.* **2021**, *18*, 100376. [CrossRef]
212. Lakamanahalli, A.; Hudedmani, M.; Shweta, G.M.; Hoskeri, P.; Mathad, S. Metamaterials: A Comprehensive Review of Design and Applications. *Int. J. Adv. Sci. Eng.* **2024**, *11*, 3816. [CrossRef]
213. Lipworth, G.; Ensworth, J.; Seetharam, K.; Da, H.; Lee, J.S.; Schmalenberg, P.; Nomura, T.; Reynolds, M.S.; Smith, D.R.; Urzhumov, Y. Magnetic metamaterial superlens for increased range wireless power transfer. *Sci. Rep.* **2014**, *4*, 3642. [CrossRef] [PubMed]
214. Mezache, Z.; Hafdi, Z.; Tao, J. Design of a novel graphene buzzle metamaterial refractometer for sensing of cancerous cells in the terahertz regime. *Optik* **2023**, *287*, 171170. [CrossRef]
215. Santos, W.F.D.; Rodrigues, A.S.L.; Lopes, I.A.R.; Pires, F.M.A.; Proença, S.P.B.; Silveira, Z.C. Analysis of a Novel 3D-Printed Mechanical Metamaterial with Tension-Induced Undulation: Experimental and Numerical Investigations. Available online: https://ouci.dntb.gov.ua/en/works/4zX8aO3r/ (accessed on 5 April 2025).
216. Dadouche, N.; Mezache, Z.; Tao, J.; Ali, E.; Alsharef, M.; Alwabli, A.; Jaffar, A.; Alzahrani, A.; Berazguia, A. Design and Fabrication of a Novel Corona-Shaped Metamaterial Biosensor for Cancer Cell Detection. *Micromachines* **2023**, *14*, 2114. [CrossRef] [PubMed]
217. Jagadeesan, V.; Venkatachalam, D.; Vinod, V.M.; Loganathan, A.K.; Muthusamy, S.; Krishnamoorthy, M.; Sadasivuni, K.K.; Geetha, M. Design and development of a new metamaterial sensor-based Minkowski fractal antenna for medical imaging. *Appl. Phys. A Mater. Sci. Process.* **2023**, *129*, 391. [CrossRef]
218. Yue, S.; Xia, W.; Yanan, L.; Mengjun, W. A Midfield Wireless Energy Transmission Antenna Based on Metamaterial for Implanted Capsule Endoscope. In Proceedings of the 2021 13th Global Symposium on Millimeter-Waves & Terahertz (GSMM), Nanjing, China, 23–26 May 2021; pp. 1–3.
219. Ren, Z.; Chang, Y.; Ma, Y.; Shih, K.; Dong, B.; Lee, C. *Leveraging of MEMS Technologies for Optical Metamaterials Applications*; Wiley-VCH Verlag: Weinheim, Germany, 2020.
220. Dhama, R.; Yan, B.; Palego, C.; Wang, Z. Super-resolution imaging by dielectric superlenses: Tio2 metamaterial superlens versus batio3 superlens. *Photonics* **2021**, *8*, 222. [CrossRef]
221. Mollaei, M.S.M.; Simovski, C. Dual-metasurface superlens: A comprehensive study. *Phys. Rev. B* **2019**, *100*, 205426. [CrossRef]
222. Smith, D. Applied physics. How to build a superlens. *Science* **2005**, *308*, 502–503. [CrossRef]
223. Raghavan, S.; Rajeshkumar, V. An Overview of Metamaterials in Biomedical Applications. In Proceedings of the Progress in Electromagnetics Research Symposium, Taipei, Taiwan, 25–28 March 2013.
224. Veerabagu, U.; Palza, H.; Quero, F. *Review: Auxetic Polymer-Based Mechanical Metamaterials for Biomedical Applications*; American Chemical Society: Washington, DC, USA, 2022.
225. Sabban, A. Small New Wearable Metamaterials Antennas for IOT, Medical and 5G Applications. In Proceedings of the 2020 14th European Conference on Antennas and Propagation (EuCAP), Copenhagen, Denmark, 15–20 March 2020.
226. Yu, X.; Zhou, J.; Liang, H.; Jiang, Z.; Wu, L. Mechanical metamaterials associated with stiffness, rigidity and compressibility: A brief review. *Prog. Mater. Sci.* **2018**, *94*, 114–173. [CrossRef]
227. Shirzad, M.; Zolfagharian, A.; Bodaghi, M.; Nam, S.Y. Auxetic metamaterials for bone-implanted medical devices: Recent advances and new perspectives. *Eur. J. Mech. A/Solids* **2023**, *98*, 104905. [CrossRef]
228. Khan, S.A.; Khan, N.Z.; Xie, Y.; Abbas, M.T.; Rauf, M.; Mehmood, I.; Runowski, M.; Agathopoulos, S.; Zhu, J. *Optical Sensing by Metamaterials and Metasurfaces: From Physics to Biomolecule Detection*; John Wiley and Sons: Hoboken, NJ, USA, 2022.
229. Xu, J.; Cai, H.; Wu, Z.; Li, X.; Tian, C.; Ao, Z.; Niu, V.C.; Xiao, X.; Jiang, L.; Khodoun, M.; et al. Acoustic metamaterials-driven transdermal drug delivery for rapid and on-demand management of acute disease. *Nat. Commun.* **2023**, *14*, 869. [CrossRef]
230. Gao, N.; Zhang, Z.; Deng, J.; Guo, X.; Cheng, B.; Hou, H. Acoustic Metamaterials for Noise Reduction: A Review. *Adv. Mater. Technol.* **2022**, *7*, 2100698. [CrossRef]
231. Sabban, A. New Wideband Meta Materials Printed Antennas for Medical Applications. *Int. J. Adv. Med. Sci.* **2015**, *3*, 13–24. [CrossRef]
232. Kasban, H.; El-Bendary, M.A.M.; Salama, D.H. A Comparative Study of Medical Imaging Techniques. *Int. J. Inf. Sci. Intell. Syst.* **2015**, *4*, 37–58.
233. Fang, W.; Lv, X.; Ma, Z.; Liu, J.; Pei, W.; Geng, Z. A Flexible Terahertz Metamaterial Biosensor for Cancer Cell Growth and Migration Detection. *Micromachines* **2022**, *13*, 631. [CrossRef]

234. Chen, J.; Chen, J.; Wang, H.; He, L.; Huang, B.; Dadbakhsh, S.; Bartolo, P. Fabrication and development of mechanical metamaterials via additive manufacturing for biomedical applications: A review. *Int. J. Extrem. Manuf.* **2025**, *7*, 012001. [CrossRef]
235. Vogiatzis, P.; Chen, S.; Wang, X.; Li, T.; Wang, L. Topology optimization of multi-material negative Poisson's ratio metamaterials using a reconciled level set method. *CAD Comput. Aided Des.* **2017**, *83*, 15–32. [CrossRef]
236. Kirillova, A.; Maxson, R.; Stoychev, G.; Gomillion, C.T.; Ionov, L. 4D Biofabrication Using Shape-Morphing Hydrogels. *Adv. Mater.* **2017**, *29*, 1703443. [CrossRef]
237. Jaafar, A.; Hecker, C.; Árki, P.; Joseph, Y. Sol-gel derived hydroxyapatite coatings for titanium implants: A review. *Bioengineering* **2020**, *7*, 127. [CrossRef]
238. Lecina-Tejero, Ó.; Pérez, M.Á.; García-Gareta, E.; Borau, C. The rise of mechanical metamaterials: Auxetic constructs for skin wound healing. *J. Tissue Eng.* **2023**, *14*, 20417314231177838. [CrossRef]
239. Wang, H.; Lyu, Y.; Bosiakov, S.; Zhu, H.; Ren, Y. A review on the mechanical metamaterials and their applications in the field of biomedical engineering. *Front. Media* **2023**, *10*, 1273961. [CrossRef]
240. Liu, R.P.; Zhao, Z.Y.; Ji, C.L.; Zhou, T. Metamaterials beyond negative refractive index: Applications in telecommunication and sensing. *Sci. China Technol. Sci.* **2016**, *59*, 1007–1011. [CrossRef]
241. Mahmud, S.; Nezaratizadeh, A.; Satriya, A.B.; Yoon, Y.K.; Ho, J.S.; Khalifa, A. Harnessing metamaterials for efficient wireless power transfer for implantable medical devices. *Bioelectron. Med.* **2024**, *10*, 7. [CrossRef] [PubMed]
242. Yang, Y. Overview of the Current State of Research on Metamaterials in Biomedicine. *BIO Web Conf.* **2024**, *142*, 03020. [CrossRef]
243. Zeenat, L.; Zolfagharian, A.; Sriya, Y.; Sasikumar, S.; Bodaghi, M.; Pati, F. 4D Printing for Vascular Tissue Engineering: Progress and Challenges. *Adv. Mater. Technol.* **2023**, *8*, 2300200. [CrossRef]
244. Li, N.; Liu, L.; Liu, Y.; Leng, J. Metamaterial-Based Electronic Skin with Conformality and Multisensory Integration. *Adv. Funct. Mater.* **2024**, *34*, 2406789. [CrossRef]
245. Algarin, J.M.; Lopez, M.A.; Freire, M.J.; Marques, R. Signal-to-noise ratio evaluation in resonant ring metamaterial lenses for MRI applications. *New J. Phys.* **2011**, *13*, 115006. [CrossRef]
246. Li, B.; Xie, R.; Sun, Z.; Shao, X.; Lian, Y.; Guo, H.; You, R.; You, Z.; Zhao, X. Nonlinear metamaterials enhanced surface coil array for parallel magnetic resonance imaging. *Nat. Commun.* **2024**, *15*, 7949. [CrossRef]
247. Freire, M.J.; Jelinek, L.; Marques, R.; Lapine, M. On the applications of μr=-1 metamaterial lenses for magnetic resonance imaging. *J. Magn. Reson.* **2010**, *203*, 81–90. [CrossRef]
248. Duan, G.; Zhao, X.; Anderson, S.W.; Zhang, X. Boosting magnetic resonance imaging signal-to-noise ratio using magnetic metamaterials. *Commun. Phys.* **2019**, *2*, 35. [CrossRef]
249. Hurshkainen, A.; Nikulin, A.; Georget, E.; Larrat, B.; Berrahou, D.; Neves, A.L.; Sabouroux, P.; Enoch, S.; Melchakova, I.; Belov, P.; et al. A Novel Metamaterial-Inspired RF-coil for Preclinical Dual-Nuclei MRI. *Sci. Rep.* **2018**, *8*, 9190. [CrossRef]
250. Jiang, Y.; Liu, Z.; Wang, C.; Chen, X. Heterogeneous Strain Distribution of Elastomer Substrates to Enhance the Sensitivity of Stretchable Strain Sensors. *Acc. Chem. Res.* **2019**, *52*, 82–90. [CrossRef] [PubMed]
251. Jiang, Y.; Liu, Z.; Matsuhisa, N.; Qi, D.; Leow, W.R.; Yang, H.; Yu, J.; Chen, G.; Liu, Y.; Wan, C.; et al. Auxetic Mechanical Metamaterials to Enhance Sensitivity of Stretchable Strain Sensors. *Adv. Mater.* **2018**, *30*, 1706589. [CrossRef]
252. Luo, J.; Lu, W.; Jiao, P.; Jang, D.; Barri, K.; Wang, J.; Meng, W.; Kumar, R.P.; Agarwal, N.; Hamilton, D.K.; et al. Wireless electronic-free mechanical metamaterial implants. *Mater. Today* **2025**, *83*, 145–156. [CrossRef]
253. Xiong, H.; Xie, J.Y.; Liu, Y.J.; Wang, B.X.; Xiao, D.P.; Zhang, H.Q. Microwave Hyperthermia Technology Based on Near-Field Focused Metasurfaces: Design and Implementation. *Adv. Funct. Mater.* **2024**, *35*, 2411842. [CrossRef]
254. Hassanpour, M.; Hassanpour, M.; Rezaie, M.; Khezripour, S.; Faruque, M.R.I.; Khandaker, M.U. The application of graphene/h-BN metamaterial in medical linear accelerators for reducing neutron leakage in the treatment room. *Phys. Eng. Sci. Med.* **2023**, *46*, 1023–1032. [CrossRef]
255. Jaffar, N.A.; Buniyamin, N.; Lias, K. An overview of available metamaterial-based antenna for non-invasive hyperthermia cancer treatment. *Indones. J. Electr. Eng. Comput. Sci.* **2019**, *14*, 697–705. [CrossRef]
256. Leggio, L.; Varona, O.D.; Dadrasnia, E. A Comparison between Different Schemes of Microwave Cancer Hyperthermia Treatment by Means of Left-Handed Metamaterial Lenses. *Prog. Electromagn. Res.-Pier* **2015**, *150*, 73–87. [CrossRef]
257. Osipkov, A.; Makeev, M.; Garsiya, E.; Filyaev, A.; Sinyagaeva, K.; Kirillov, D.; Ryzhenko, D.; Yurkov, G. Radio-shielding metamaterials transparent in the visible spectrum: Approaches to creation. *IOP Conf. Ser. Mater. Sci. Eng.* **2021**, *1060*, 012007. [CrossRef]
258. Anwar, U.; Arslan, T.; Hussain, A.; Lomax, P. Next Generation Cognition-Aware Hearing Aid Devices With Microwave Sensors: Opportunities and Challenges. *IEEE Access* **2022**, *10*, 82214–82235. [CrossRef]

crystals

MDPI

Article

A Long-Wave Infrared Circularly Polarized Photodetector Based on an Array of Trapezoidal Silicon Pillars

Bo Cheng [1,2], Yuxiao Zou [3], Taohua Liang [1,2,*], Ansheng Ye [1,2], Kunpeng Zhai [4,*] and Longfeng Lv [5,*]

1 Postdoctoral Innovation Practice Base, Chengdu Polytechnic, 83 Tianyi Street, Chengdu 610041, China; chengbo9610@semi.ac.cn (B.C.); cduyas@163.com (A.Y.)
2 Sichuan Provincial Engineering Research Center of Thermoelectric Materials and Devices, Chengdu 610041, China
3 Kunming Institute of Physics, Kunming 650223, China
4 Institute of Intelligent Photonics, Nankai University, Tianjin 300071, China
5 Institute of Semiconductors, Chinese Academy of Sciences, Beijing 100083, China
* Correspondence: lthua222@163.com (T.L.); kpzhai@nankai.edu.cn (K.Z.); lflv@semi.ac.cn (L.L.)

Abstract: Integrating metasurface-based polarizing filters atop photodetectors enables the expansion of detection capabilities from intensity to polarization, offering significant potential for applications requiring high-precision discrimination in scientific, industrial, and defense sectors. However, such metasurfaces often introduce optical efficiency losses. Here, we present a long-wave infrared (8.6 μm) circularly polarized photodetector capable of direct chiral discrimination, eliminating the need for additional optical components. The polarization selectivity arises from Guided-Mode resonances (GMRs) excited by two horizontally offset right-trapezoidal unit cells within a chiral metasurface. This design exhibits a pronounced transmittance contrast (~100%) between left circularly polarized light (LCP) and right circularly polarized light (RCP) while maintaining fabrication simplicity via a conventional single-step lithographic process. The proposed detector is expected to achieve high-dimensional physical characterization by resolving polarization-encoded vectorial information, demonstrating enhanced performance in complex environments.

Keywords: metasurfaces; circularly polarized photodetector; GMRs

1. Introduction

Conventional photodetectors [1] are fundamentally limited to measuring light intensity and can only provide scalar information about optical signals. Although this is sufficient for many applications, in critical fields such as quantum communication [2], remote sensing [3], and biomedical imaging [4], it is necessary to obtain the circular polarization state of light—this chiral vector property carries important information about the interaction of light with matter. Traditional methods for circular polarization detection rely on cascaded optical elements [5] (such as waveplates and linear polarizers), which inevitably lead to system complexity, alignment difficulties, and significant optical losses.

Metasurfaces [6–10] have emerged as a promising platform for manipulating optical signals, enabling flexible control of amplitude, phase, and polarization at subwavelength scales. Chiral metasurfaces [11–15], a subset of metasurfaces, can generate differential electromagnetic responses to circularly polarized light (such as selective absorption, phase modulation, or transmission differences), thereby actively regulating the spin degree of freedom of light. Chiral metasurfaces are generally characterized by two metrics: circular dichroism [16–18] and circular polarization extinction ratio [19–21], which correspond to

 https://doi.org/10.3390/cryst15110993

the difference and ratio of the signals of left and right circularly polarized light, respectively. Chiral metasurfaces offer an intriguing lightweight alternative to cascaded optical components, as their breaking of mirror symmetry [22–25] allows for the direct discrimination of circularly polarized light without the need for auxiliary optical devices.

In this paper, we designed a long-wave infrared circularly polarized photodetector (CPPD) with an efficiency close to 100% based on the COMSOL 5.6 simulation platform. Firstly, we introduced the structural composition, material information and optical model configuration of the CPPD, analyzed the influence of the geometric changes in the unit cell structure on circular dichroism, then dissected the characteristics of the electric field distribution map of the chiral metasurface filter and the impact of the thickness of the support layer *hs* on the coupling between the metasurface and the detector, and finally demonstrated the possible impact of process errors on this device.

2. Materials and Methods

Figure 1a illustrates the three-dimensional structure of the CPPD, which consists of a silicon metasurface on top and a type-II superlattice photodetector [26,27] operating at a wavelength of 8.6 μm at the bottom. As shown in Figure 1b, the silicon metasurface is a periodically arranged array, and its unit cell is composed of two right-angled trapezoidal columns. Figure 1c presents the information on the cross-sectional view of the silicon metasurface in the XY plane, revealing that the lower trapezoid can be obtained by rotating the upper trapezoid 180 degrees counterclockwise. The long-wave infrared type-II superlattice photodetector, from bottom to top, consists of a 350 μm thick GaSb substrate, a 0.67 μm thick GaSb buffer layer, a 0.1 μm thick p-type doped GaSb lower contact layer, approximately 400 periods of InAs/GaSb absorption layers, a 0.16 μm thick AlGaSb barrier layer, a 0.37 μm thick n-type doped InAs upper contact layer, and a 3.1 μm thick MgF_2 antireflection layer. The optical refractive index of the absorption region in the type-II superlattice is estimated by the effective medium theory (EMT) [28,29], and the imaginary part of the refractive index at a wavelength of 8.6 μm is approximately 0.09, as shown in Figure 1d. The absorption rate of the CPPD is obtained by the simulation COMSOL Multiphysics, which is equal to 1 minus the reflectance minus the transmittance. Periodic boundary conditions (PBC) are chosen for the four sides of the optical simulation model, while perfect matching layers are added to the upper and lower end faces to truncate the electromagnetic field. The polarization characteristics of the incident light are controlled by the incident port. The reflectance and transmittance are, respectively, the area integrals of the Poynting vector at the reflection port and the incident port. More details of the simulation model and process can be found in the Supplementary Materials.

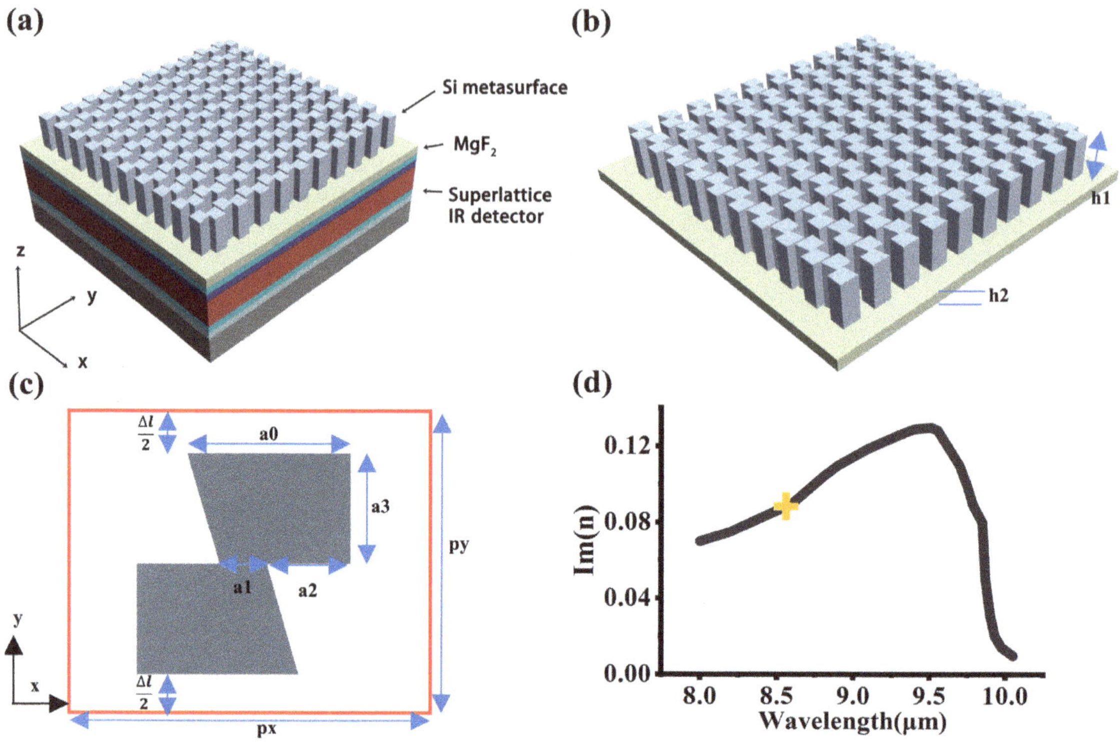

Figure 1. The structure and materials of micro-devices. (**a**) A circularly polarized type-II superlattice photodetector operating at a wavelength of 8.6 μm; (**b**) Three-dimensional structure of chiral metasurface. The h1 = 5.59 μm, h2 = 3.1 μm. (**c**) The XY cross-sectional view of the unit cell of the metasurface. The px = py = 4 μm, a0 = 2.19 μm, a1 = 0.64 μm, a2 = 1.17 μm, a3 = 1.88 μm. (**d**) The imaginary part of the optical refractive index of the absorption region in a type-II superlattice photodetector. The orange cross cursor corresponds to 8.6 μm, which is a characteristic wavelength for monitoring the concentration and distribution of ozone in the atmosphere [30,31].

3. Results

Figure 2a shows the transmission spectrum of the chiral metasurface in the wavelength range from 8.0 μm to 9.2 μm. In case of the LCP incidence, the transmission first increases, then decreases, and then increases again, with the maximum transmission of 0.97 corresponding to the operating wavelength of 8.6 μm. For RCP, there is a significant attenuation at the wavelength of 8.6 μm, and the transmission is already close to 0. Clearly, near the wavelength of 8.6 μm, the chiral metasurface exhibits distinct transmission characteristics for different chiral circularly polarized light. As shown in Figure 2b, the circular polarization extinction ratio (CPER) spectrum of the silicon metasurface filter shows the characteristics of a Dirac function. At the wavelength near 8.6 μm, CPER is a sharp peak, and on both sides far from the 8.6 μm wavelength, the value of CPER rapidly approaches 0. Here, CPER $= 10 \times \log(T_{\mathrm{LCP}}/T_{\mathrm{RCP}})$, where T_{LCP} and T_{RCP} are the transmission of the chiral metasurface for LCP and RCP, respectively.

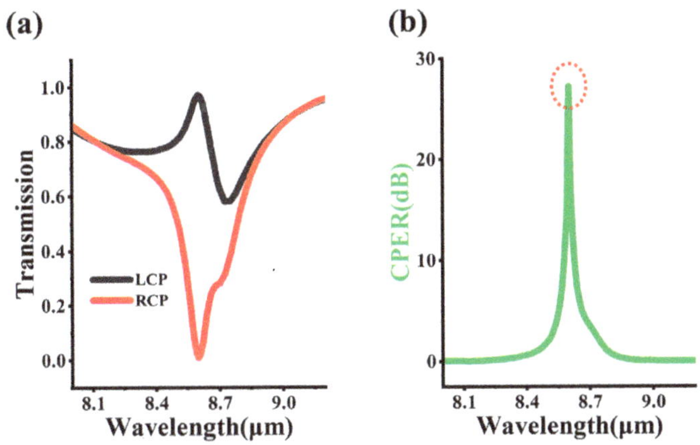

Figure 2. Transmission characteristics of the chiral metasurface. (**a**) Transmission spectra of circularly polarized light; (**b**) CPER spectra. The red dotted circle corresponds to the CPER at the operating wavelength.

3.1. The Influence of Geometric Structures in Metasurfaces

Generally, the geometric structure of the unit cell of the metasurface can directly determine the peak position, line shape and intensity of the circular dichroism (CD) spectrum, and its asymmetric parameters (such as displacement Δx, tilt angle θ) have a close mapping relationship with the CD value. Figure 3a–d, respectively, show the influence of the geometric changes in the unit cell of the chiral metasurface on the CD spectrum, where $CD = T_{LCP} - T_{RCP}$. Figure 3a shows the case where the unit cell of the metasurface is two aligned rectangles, which is a typical C_2 symmetric structure. Figure 3e shows the CD spectrum of the metasurface corresponding to the structure in Figure 3a, which is a constant zero, reflecting the fact that the unit cell of the metasurface is a non-chiral symmetric structure. Compared with Figure 3a,b introduces a horizontal displacement dx, which makes the unit cell of the metasurface transition from a non-chiral symmetric structure to a chiral symmetric structure. Figure 3f shows the CD spectrum of the geometric structure in Figure 3b, which presents a typical double-hump function spectrum shape, with the maximum CD peak value being approximately 0.2. In Figure 3c, triangles are added to both sides of the rectangle, introducing the concept of a tilted structure. As shown in Figure 3g, a huge CD peak (>0.8) appears at a wavelength of 8.6 μm, indicating that the metasurface has acquired significant chiral symmetry after undergoing dual changes in displacement and tilt angle. The structures in Figure 3c,d satisfy the mirror symmetry relationship about the YZ plane, which means that the metasurface's control over the transmission of chiral beams has been reversed; that is, the CD values at each wavelength in Figure 3g,h are opposite to each other.

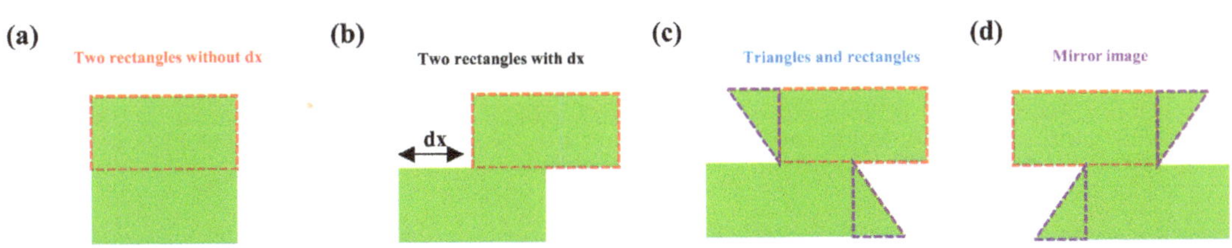

Figure 3. *Cont.*

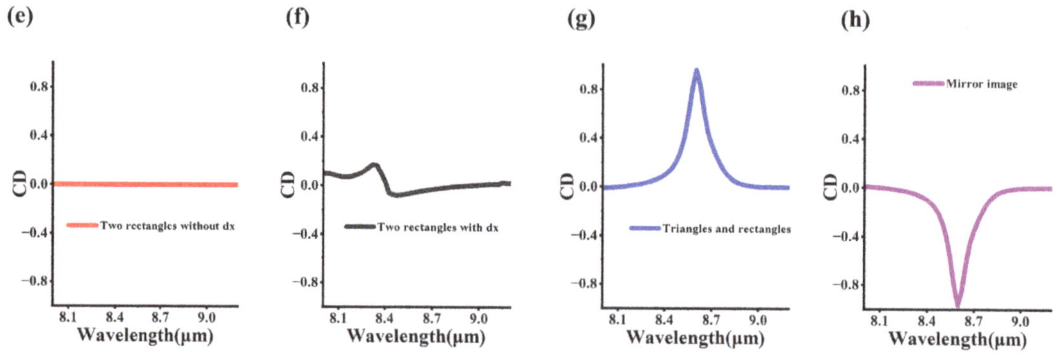

Figure 3. The influence of geometric operations. (**a**) Two aligned rectangles; (**b**) Two rectangles with a horizontal dislocation dx; (**c**) The combination of the rectangle and the triangle; (**d**) The mirror image pattern of Figure 3c in the YZ plane. The colored dotted lines correspond to the boundaries of the structures located above the long strips in the unit cell; (**e**) The CD spectrum corresponding to Figure 3a; (**f**) The CD spectrum corresponding to Figure 3b; (**g**) The CD spectrum corresponding to Figure 3c; (**h**) The CD spectrum corresponding to Figure 3d.

3.2. Analysis of the Electric Field Distribution of Metasurfaces

Observing the near-field intensity distribution of metasurfaces is also an effective means to analyze the optical resonance modes. Figure 4a shows the electric field intensity distribution of the metasurface on different XY cross-sections, all sharing the same color scale. All six sub-figures indicate a strong electric field localization phenomenon in the air gap between adjacent unit cells along the Y-axis, which is represented by the red areas in the images. Additionally, as shown in Figure 4b,c, there is a clear linear relationship between the length of the air gap, Δl, and the incident wavelength, which exhibits the typical characteristics of the GMRs. The red dashed circles and black boxes, respectively, indicate the cases corresponding to the metasurface structure in Figure 1, which simultaneously correspond to the extremely high LCP transmission and extremely low RCP transmission. Figure 4d shows the electric field localization contrast (ELC) of the chiral metasurface for different chiral CPs, which quantitatively measures the concentration degree of the electromagnetic field energy in Figure 4a. The maximum ELC corresponds to a wavelength of 8.6 μm, indicating a direct correlation between the resonance wavelength and the distribution of electric field hotspots. Here, ELC $= Q_{LCP} - Q_{RCP}$, $Q_{LCP} = \int \frac{|E_L|}{E_0} dV / I_{max}$, $Q_{RCP} = \int \frac{|E_R|}{E_0} dV / I_{max}$, where dV corresponds to the volume occupied by the metasurface. I_{max} is the normalization term in the denominator and is the maximum value of $\int \frac{|E_L|}{E_0} dV$. E_0 corresponds to the electric field intensity of the incident light. E_L and E_R, respectively, correspond to the electric field intensities when the incident light is LCP and RCP.

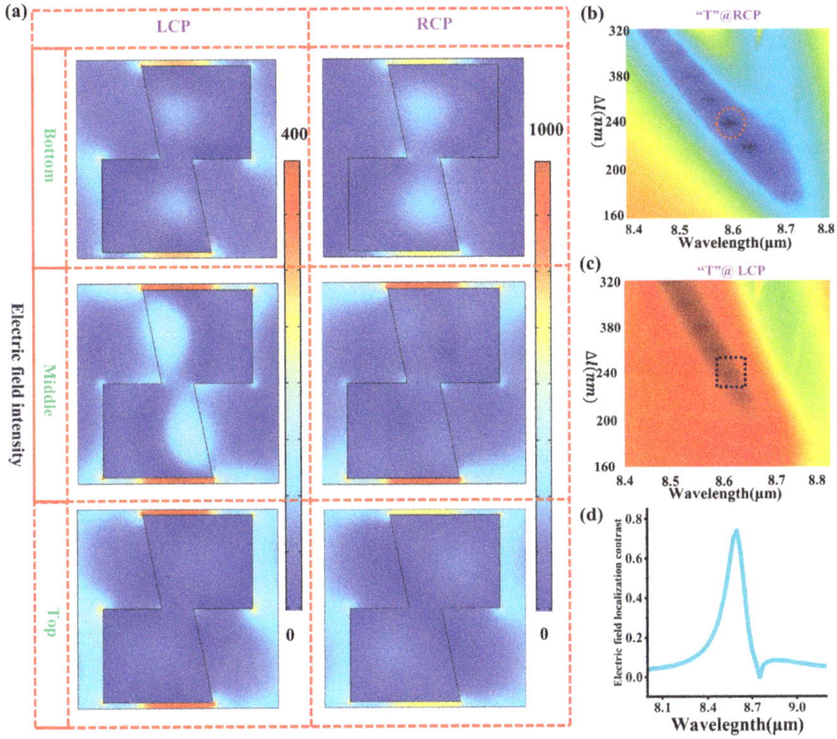

Figure 4. The distribution of the electric field of the metasurface and the GMRs. (**a**) Electric field distribution on the XY cross-section of the metasurface. "Bottom", "Middle", and "Top" correspond to the positions of the XY cross-section at the bottom, middle, and top of the metasurface, respectively; (**b**,**c**) The influence of the air gap length Δl on the transmission spectrum; (**d**) The difference in the localization of the electric field in the LCP and RCP cases. The dashed circles and boxes correspond to the transmittance at the operating wavelength.

3.3. The Influence of the Thickness of the Support Layer (hs)

In the previous sections, we focused on analyzing the transmission characteristics of chiral metasurfaces for different circularly polarized lights. In this subsection, we mainly discuss the influence of the thickness *hs* of the support layer (MgF2) between the chiral metasurface and the type-II superlattice photodetector on chirality. One fact must be noted, which is that we have pre-assumed the metasurface and the photodetector to be two independent sub-modules when designing the CPPD, so *hs* must be carefully scanned to ensure that the coupling effect between them is not a negative feedback. Figure 5a depicts the impact of *hs* on the CPPD at a wavelength of 8.6 μm. The absorption spectrum of LCP has a typical fluctuating characteristic, while the absorption spectrum of RCP first decreases and then remains basically unchanged. CPER is also a fluctuating curve, with the maximum CPER value corresponding to *hs* of 3.1 μm. Here, CPER $= 10 \times \log(A_{LCP}/\ A_{RCP})$, where A_{LCP} and A_{RCP} are the absorption rate of the chiral metasurface filter for LCP and RCP, respectively. Figure 5b shows the absorption spectrum and CPER spectrum of the CPPD when *hs* = 3.1 μm. The maximum LCP absorption rate (≈0.33) and CPER value (≈28 dB) both occur at the operating wavelength of 8.6 μm. Figure 5c–f show the electric field distribution in the detector region under different CP incident conditions. It can be found that a large amount of electric field energy from the upper region penetrates into the absorption layer corresponding to LCP. However, in the case of RCP incidence, the upper metasurface acts like a perfect reflector, and almost no free photons can enter the absorption region, which is represented by the blue color indicating low electromagnetic energy in the picture.

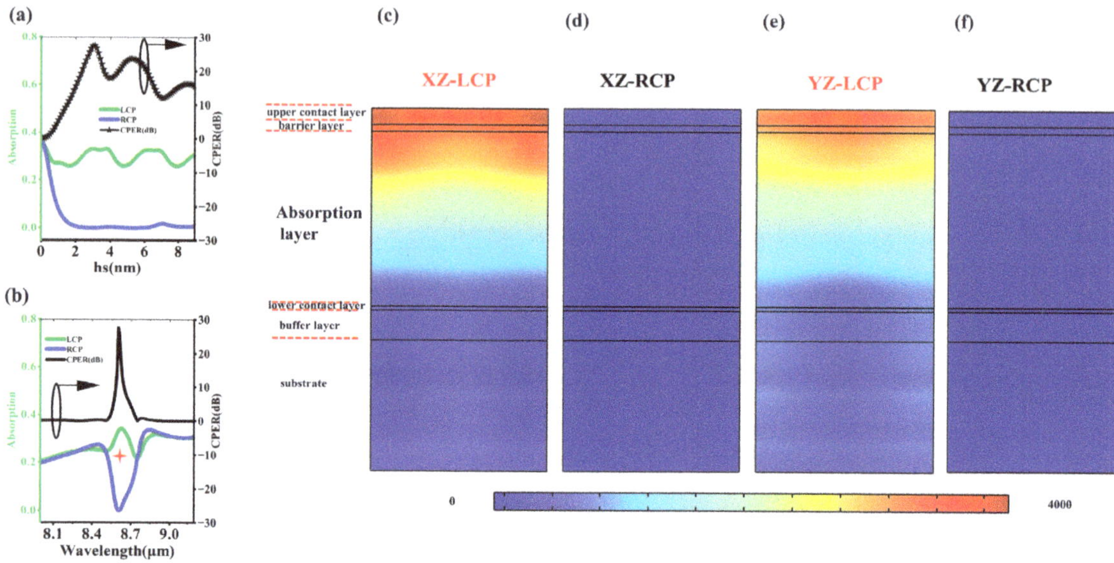

Figure 5. (**a**) The influence of the thickness of the supporting layer *hs* on the absorption of the left circularly polarized photodetector at an operating wavelength of 8.6 μm; (**b**) The absorption spectrum and CPER spectrum of the polarization detector corresponding to the optimal *hs*, The plus sign corresponds to the position of the peak; (**c**) The electric field distribution diagram of the CPPD under LCP incidence in the XZ cross-section; (**d**) The electric field distribution diagram of the CPPD under RCP incidence in the XZ cross-section; (**e**) The electric field distribution diagram of the CPPD under LCP incidence in the YZ cross-section; (**f**) The electric field distribution diagram of the CPPD under RCP incidence in the YZ cross-section.

3.4. Analysis of Potential Process Errors

Since the chirality of the polarization detector is determined by the nanostructure pattern of the metasurface, the influence of line errors cannot be ignored. In semiconductor processing, it is impossible to achieve an ideal rectangle, and instead, a circular arc replaces the right angle of the rectangle. This is a result of the combined effects of photolithography diffraction, microloading during etching, and the natural tendency of surface energy minimization. Figure 6a shows the influence of the radius of the inscribed circle in the silicon metasurface on the polarization detector. When the radius is less than 100 nm, the absorption rates of LCP and RCP remain almost unchanged. When the radius is greater than 100 nm, the two absorption rates approach each other as the radius increases. Additionally, over-etching and under-etching phenomena are often encountered during semiconductor etching, and the resulting etching depth errors also require additional attention. The inset in Figure 6b shows the over-etching phenomenon, where not only is the silicon completely etched away, but also the underlying substrate material is patterned. As shown in Figure 6b, the change caused by the etching error Δh is weak. However, a completely different phenomenon occurs in Figure 6c. For LCP, when the current etching error Δh is within the range of 0 nm to 60 nm, the absorption rate decreases significantly as Δh increases. When Δh is greater than 60 nm, the absorption rate remains almost unchanged. Moreover, for RCP, the absorption rate increases as Δh increases. When

Δh equals 150 nm, the absorption rates of RCP and LCP are the same, indicating that the detector has lost its chiral detection capability. Therefore, it is very necessary to intentionally over-etch for a few more seconds during etching to guarantee that the etching process unequivocally reaches the over-etching state. Table 1 presents the performance of some chiral metasurfaces published in the recent two years. It can be found that the CD and CPER of the metasurfaces in the manuscripts are both excellent. Particularly for the CPER index, our work has basically led by one order of magnitude.

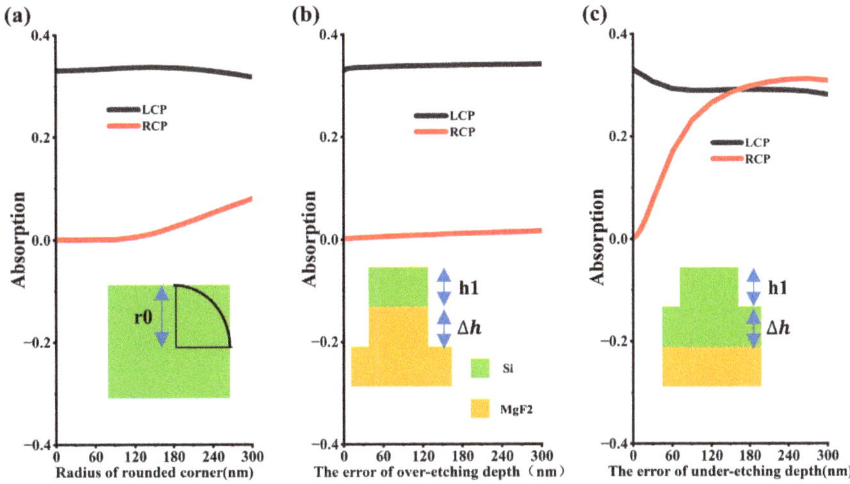

Figure 6. The influence of potential process errors. (**a**) Radius of the fillet. (**b**) Error of over-etching depth. (**c**) Error of under-etching depth. Green represents silicon.

Table 1. A comparison of the performance of chiral metasurfaces.

Structure	Publication	Wavelength (μm)	CD	CPER (dB)
π-shaped [32]	2024	1.55, 2.55	0.7, 0.75	~10, ~10
Monoclinic Lattices [33]	2024	1.6	0.3	/
Broken circle [34]	2025	1.35	0.17	/
Graphene and silicon [35]	2024	250	0.9	9
Double Si columns [36]	2024	1.01	~0.9	6
Swastika-shaped [37]	2024	0.53	0.26	3
Double-layered gold [38]	2025	1.25	0.55	~8
This work		8.6	~1	27

4. Discussion

We have developed a novel CPPD capable of direct chiral discrimination in free-space optical systems, comprising a silicon-based chiral metasurface integrated with a type-II superlattice photodetector. By precisely engineering the air gap dimension (Δl) between adjacent meta-atoms, our device achieves near 100% CD at the operational wavelength of 8.6 μm through excitation of optimized GMRs modes. This CPPD scheme provides access to the chirality of photons—a critical light parameter that remains inaccessible to conventional intensity-based photodetectors, thereby opening new possibilities in quantum optics and interdisciplinary applications. Future research directions will focus on two key aspects, namely (1) bandwidth enhancement to enable spectrally broadband operation and (2) integration of chiral metasurfaces with dual-color or multicolor detection systems, to simultaneously achieve multi photon acquisition and polarization-resolved measurements.

Supplementary Materials: The following supporting information can be downloaded at https://www.mdpi.com/article/10.3390/cryst15110993/s1. Figure S1: (a): XZ cross-sectional view of the COMSOL model of the chiral metasurface. (b): XZ cross-sectional view of the COMSOL model of the circularly polarized photodetector. Figure S2: Basic Structure of TEC.

Author Contributions: Conceptualization, B.C. and Y.Z.; methodology, A.Y.; software, B.C.; validation, B.C., Y.Z. and K.Z.; formal analysis, A.Y.; investigation, T.L.; resources, K.Z.; data curation, L.L.; writing—original draft preparation, B.C.; writing—review and editing, K.Z.; visualization, K.Z.; supervision, K.Z.; project administration, L.L.; funding acquisition, K.Z. All authors have read and agreed to the published version of the manuscript.

Funding: This research was funded by the Natural Science Foundation of Tianjin, grant number No. 24JCQNJC01880; This research was funded by the Research on Optoelectronic Fusion Technology, ZB253015; This research was funded by the Young Elite Scientists Sponsorship Program of Cast, grant number No. YESS20240775; This research was funded by the Strategic Priority Research Program of the Chinese Academy of Sciences (grant no. XDB0980000).

Data Availability Statement: The raw data supporting the conclusions of this article will be made available by the authors on request.

Acknowledgments: The authors are grateful to Jietao Liu, Hanxiao Shao and GuoFeng Song of Institute of Semiconductors, Chinese Academy of Sciences.

Conflicts of Interest: The authors declare no conflicts of interest.

References

1. Smith, W.H.; Born, J.; Cochran, W.D.; Gelfand, J. Spectroscopic photoelectric imaging Fabry-Perot interferometer—Its development and preliminary observational results. *Appl. Opt.* **1976**, *15*, 717–724. [CrossRef]
2. Pathak, A. *Elements of Quantum Computation and Quantum Communication*; Taylor & Francis: Abingdon, UK, 2013.
3. King, M.D.; Kaufman, Y.J.; Tanré, D.; Nakajima, T. Remote sensing of tropospheric aerosols from space: Past, future and present. *Bull. Am. Meteorol. Soc.* **1999**, *80*, 2229–2259. [CrossRef]
4. Rudolph, W.; Kempe, M. Trends in optical biomedical imaging. *Opt. Acta Int. J. Opt.* **1997**, *44*, 1617–1642. [CrossRef]
5. Sambles, J.R. Polarized Light in Optics and Spectroscopy. *Opt. Acta Int. J. Opt.* **1991**, *38*, 1204–1205. [CrossRef]
6. Liu, Y.; Ling, X.; Yi, X.; Zhou, X.; Luo, H.; Wen, S. Realization of polarization evolution on higher-order Poincare sphere with metasurface. *Appl. Phys. Lett.* **2014**, *104*, 191110. [CrossRef]
7. Wei, Z.; Cao, Y.; Su, X.; Gong, Z.; Li, H. Highly efficient beam steering with a transparent metasurface. *Opt. Express* **2013**, *21*, 10739–10745. [CrossRef] [PubMed]
8. Arbabi, A.; Arbabi, E.; Horie, Y.; Kamali, S.M.; Faraon, A. Planar metasurface retroreflector. *Nat. Photonics* **2017**, *11*, 415–420. [CrossRef]
9. Liu, C.; Wang, H.; Zhao, R.; Lei, Y.; Dong, S.; Cai, Y.; Zhou, W.; Wang, Y.; Huang, L.; Xu, K.M. Multi-plane vectorial holography based on a height tunable metasurface fabricated by femtosecond laser direct writing. *Photonics Res.* **2024**, *12*, 8. [CrossRef]
10. Li, S.; Chen, C.; Wang, G.; Ge, S.; Zhao, J.; Ming, X.; Zhao, W.; Li, T.; Zhang, W. Metasurface Polarization Optics: Phase Manipulation for Arbitrary Polarization Conversion Condition. *Phys. Rev. Lett.* **2025**, *134*, 023803. [CrossRef] [PubMed]
11. Li, Z.; Liu, W.; Cheng, H.; Chen, S.; Tian, J. Tunable dual-band asymmetric transmission for circularly polarized waves with graphene planar chiral metasurfaces. *Opt. Lett.* **2016**, *41*, 3142–3145. [CrossRef] [PubMed]
12. Ma, C.; Yu, P.; Jing, Z.; Zhu, Y.; Li, P.; Wang, W.; Xu, H.; Zhang, Y.; Pan, L.; Choi, T.-Y.; et al. Circular polarization-selective optical, photothermal, and optofluidic effects in chiral metasurfaces. *Photonics Res.* **2024**, *12*, 331. [CrossRef]
13. Jang, J.; Park, M.; Park, Y. Non-interleaved chiral metasurfaces and neural networks enhance the spatial resolution of polarimetry. *Light Sci. Appl.* **2024**, *13*, 53. [CrossRef]
14. Wang, C.; Cheng, X.; Wang, R.; Hu, X.; Wang, C. Flexibly Designable 2D Chiral Metasurfaces with Pixelated Topological Structure Based on Machine Learning. *Laser Photonics Rev.* **2024**, *18*, 2300958. [CrossRef]
15. Gromyko, D.; Loh, J.S.; Feng, J.; Qiu, C.W.; Wu, L. Enabling All-to-Circular Polarization Up-Conversion by Nonlinear Chiral Metasurfaces with Rotational Symmetry. *Phys. Rev. Lett.* **2025**, *134*, 023804. [CrossRef]
16. Stephens, P.J.; Devlin, F.J.; Chabalowski, C.F.; Frisch, M.J. Ab Initio Calculation of Vibrational Absorption and Circular Dichroism Spectra Using Density Functional Force Fields. *J. Phys. Chem.* **1994**, *98*, 247–257. [CrossRef]
17. Greenfield, N.J.; Fasman, G.D. Computed circular dichroism spectra for the evaluation of protein conformation. *Biochemistry* **1969**, *8*, 4108. [CrossRef] [PubMed]

18. Berova, N.; Nakanishi, K.; Woody, R.W. Circular Dichroism: Principles and Applications. In *Circular Dichroism Principles Applications*; John Wiley & Sons: Hoboken, NJ, USA, 2000.
19. Zhang, R.; Zhang, Z.; Fan, Y.; Zhang, H.; Chu, J. Single-Layer Transmissive Chiral Plasma Metasurface with High Circular Polarization Extinction Ratio in Visible Wavelength. *Nanomaterials* **2023**, *13*, 813. [CrossRef]
20. Ye, Y.; Li, X.; Zhuang, F.; Chang, S.W. Homogeneous circular polarizers using a bilayered chiral metamaterial. *Appl. Phys. Lett.* **2011**, *99*, 031111. [CrossRef]
21. Cheng, B.; Zou, Y.; Song, G. Full-Stokes polarization short-wave infrared photodetector based on the doubly regulated dielectric metasurfaces. *Optik* **2025**, *322*, 172198. [CrossRef]
22. Tang, Y.; Xie, B.; Liu, H.; Zhang, Y.; Cheng, H.; Chen, S. Mirror Symmetry Broken of Sound Vortex Transmission in a Single Passive Metasurface via Phase Coupling. *Phys. Rev. Lett.* **2024**, *132*, 6. [CrossRef] [PubMed]
23. Shen, Z.; Huang, D.; Lin, X. Dual-band chirality-selective absorbing by plasmonic metasurfaces with breaking mirror and rotational symmetry. *Opt. Express* **2023**, *31*, 12. [CrossRef]
24. Petronijevic, E.; Cesca, T.; Scian, C.; Mattei, G.; Voti, R.L.; Sibilia, C.; Belardini, A. Demonstration of extrinsic chirality in self-assembled asymmetric plasmonic metasurfaces and nanohole arrays. *Sci. Rep.* **2024**, *14*, 17210. [CrossRef]
25. Levanon, N.; Indukuri, S.C.; Frydendahl, C.; Bar-David, J.; Han, Z.; Mazurski, N.; Levy, U. Angular Transmission Response of In-Plane Symmetry-Breaking Quasi-BIC All-Dielectric Metasurfaces. *ACS Photonics* **2022**, *9*, 3642–3648. [CrossRef]
26. Gordon, I.E.; Rothman, L.S.; Hargreaves, R.J.; Hashemi, R.; Karlovets, E.V.; Skinner, F.M.; Conway, E.K.; Hill, C.; Kochanov, R.V.; Tan, Y.; et al. The HITRAN2020 molecular spectroscopic database. *J. Quant. Spectrosc. Radiat. Transf.* **2022**, *277*, 107949. [CrossRef]
27. Beer, R. TES on the aura mission: Scientific objectives, measurements, and analysis overview. *IEEE Trans. Geosci. Remote Sens.* **2006**, *44*, 1102–1105. [CrossRef]
28. Brown, G.J.; Szmulowicz, F.; Linville, R.; Saxler, A.; Mahalingham, K.; Lin, C.H.; Kuo, C.H.; Hwang, W.Y. Type-II superlattice photodetector on a compliant GaAs substrate. *IEEE Photonics Technol. Lett.* **2002**, *12*, 684–686. [CrossRef]
29. Koyama, M.; Kudo, J.; Iguchi, Y.; Inada, H.; Muramatsu, Y.; Kumihashi, I.; Sano, M. Indium Gallium Arsenide/Gallium Arsenide Antimonide Type II Superlattice Photodetector That Can Utilize Night Airglow. *Sens. Mater.* **2024**, *36*, 2183–2197.
30. Wu, Y.; Li, J.; Zhang, Z.Q.; Chan, C.T. Effective medium theory for magnetodielectric composites: Beyond the long-wavelength limit. *Phys. Rev.* **2006**, *B74*, 085111, Erratum in *Phys. Rev.* **2015**, *B92*, 199901. [CrossRef]
31. Chang, L.; Ho, C.C.; Wei, H.S.; Wu, G.Y. Effective medium theory with dimensionality reduction for band structures of photonic crystals. *J. Appl. Phys.* **2007**, *101*, 2059. [CrossRef]
32. Gao, Z.; Wang, P.; Xu, Z.; Zhao, T. Chiral metasurface absorber with near-infrared excitation-induced dual circular dichroism. *Opt. Laser Technol.* **2024**, *175*, 110826. [CrossRef]
33. Toftul, I.; Tonkaev, P.; Koshelev, K.; Lai, F.; Song, Q.; Gorkunov, M.; Kivshar, Y. Chiral Dichroism in Resonant Metasurfaces with Monoclinic Lattices. *Phys. Rev. Lett.* **2024**, *133*, 216901. [CrossRef] [PubMed]
34. Sun, Y.; He, C.; Deng, Z.; Li, X.; Li, X.; Zhang, Z.; Sui, X.; Li, N.; He, W.; Chen, F. Optical chirality of all dielectric q-BIC metasurface with symmetry breaking. *Nanophotonics* **2025**, *14*, 559–569. [CrossRef] [PubMed]
35. Wang, G.; Zhang, X.; Liu, J.; Feng, H.; Li, M.; Yang, S.; Jia, Y.; Meng, H.; Gao, Y. A multifunction tunable terahertz chiral metasurface based on graphene and photosensitive silicon. *Opt. Mater.* **2024**, *150*, 115177. [CrossRef]
36. Raza, F.; Chen, R.; Ma, Y. Fano-modulated chiral metasurface for near-infrared sensing via bound states in the continuum. *J. Appl. Phys.* **2024**, *136*, 243103. [CrossRef]
37. Han, J.; Jang, H.; Lim, Y.; Kim, S.; Lee, J.; Jun, Y.C. Chiral Emission from Optical Metasurfaces and Metacavities. *Adv. Photonics Res.* **2024**, *5*, 2400060. [CrossRef]
38. Asefa, S.A.; Seong, M.; Lee, D. Design of Bilayer Crescent Chiral Metasurfaces for Enhanced Chiroptical Response. *Sensors* **2025**, *25*, 915. [CrossRef] [PubMed]

crystals

MDPI

Article

Wavelength Conversion in Photonic Crystal Fibers via Multiple Raman Redshifts and Soliton Spectral Tunneling

Mingliang Deng [1], Hui Wang [1,*] and Bing Wen [2]

1 College of Electronic Information Engineering, Loudi Vocational and Technical College, Loudi 417000, China; dengmingliangvip@163.com
2 All-Solid-State Energy Storage Materials and Devices Key Laboratory of Hunan Province, College of Information and Electronic Engineering, Hunan City University, Yiyang 413000, China; wenbing@hncu.edu.cn
* Correspondence: wanghuivvip@163.com

Abstract: The soliton spectral tunneling effect in photonic crystal fibers with three zero-dispersion wavelengths is an effective way to realize pulse wavelength conversion. However, due to the limitation of the soliton splitting mechanism, forming a tunneling soliton with larger energy and wider width by increasing the number of Raman redshifts is still a key challenge. Airyprime pulses generate polychromatic Raman solitons with small truncation coefficients, and they converge into a stable soliton after the soliton tunneling effect, which provides a new possibility to solve this problem. This paper discusses how to control the energy, width and central wavelength characteristics of the tunneling solitons generated in the photonic crystal fiber with three zero-dispersion wavelengths by adjusting the truncation coefficient α, peak power P and central wavelength λ of the initial Airyprime pulse. The results show that the smaller the truncation coefficient α, the greater the number of Raman self-frequency shifts and the greater the energy of the formed tunneling soliton. The increase in initial power P_0 will lead to an increase in tunneling soliton width. The larger initial center wavelength λ will significantly increase the width and center wavelength position of the tunneling soliton. These findings provide a theoretical basis for the application of Airyprime pulses in ultrafast optical wavelength control and new light source development.

Keywords: photonic crystal fibers; Airyprime pulse; Raman effect; tunneling effect; wavelength conversion

1. Introduction

Supercontinuum (SC) generation has long been a research hotspot in the field of optics [1]. The nonlinear dynamics of soliton tunneling offer a novel approach to enhancing SC generation efficiency [2]. When conditions permit, fundamental solitons near the central wavelength can undergo tunneling effects through a finite region of normal group velocity dispersion induced by stimulated Raman self-shifting, forming stable tunneling solitons. This region is typically sandwiched between two regions of anomalous group velocity dispersion [3]. This phenomenon shares numerous similarities with quantum tunneling in quantum mechanics; hence, it is termed Soliton Spectrum Tunneling (SST) [4,5]. The SST effect is also considered to be a way of soliton-induced dispersive wave generation, because the transmitted solitons are phase-matched with the emitted solitons. In this process, the dispersive wave is generated under the action of nonlinear effects and high-order dispersion, also known as 'Cherenkov radiation wave' [6] or 'non-soliton radiation

 https://doi.org/10.3390/cryst15110962

wave' [7]. These waves will continue to radiate energy outward, which can effectively transfer the energy of solitons to other solitons after tunneling, thereby realizing the conversion between solitons and dispersive waves [8,9]. We can utilize its short-term stability and self-dissipation characteristics to achieve temporary propagation and automatic destruction of sensitive information, thereby effectively preventing eavesdropping and interception [10]. At the same time, the SST effect can be applied to control the generation of SC, as it enables wavelength conversion of optical pulses and optimizes the energy distribution of specific wavelengths by generating high-power spectral components, thereby improving the flatness of SC [11]. Traditional optical fibers have limitations in higher-order dispersion, whereas photonic crystal fibers (PCFs) break these limitations. Due to its tunable dispersion characteristics and high nonlinearity, PCF has become the ideal medium for generating SC [12]. In 2011, S.P. Stark and other researchers successfully realized a PCF with three zero-dispersion wavelengths (ZDWs) by changing the geometric structure of the PCF and the shape and position of the air hole [13]. Compared with the double-zero-dispersion PCF, the triple-zero-dispersion PCF has a more flexible dispersion curve and a longer wavelength interval between the dispersion zeros. The first zero-dispersion wavelength is tuned to the near-infrared region, and the third zero-dispersion wavelength is tuned to the mid-infrared region, which effectively controls the broadening of the supercontinuum in both the short-wavelength and mid-infrared directions [14]. In the triple-zero-dispersion PCF, due to the simultaneous existence of positive dispersion and negative dispersion slopes, the soliton splitting process can not only generate blue-shifted dispersive waves but also generate redshifted dispersive waves at the same time [15]. The redshifted dispersive wave can excite the SST effect in the mid-infrared region, thereby increasing the width of the dispersive wave and converting the wavelength. Therefore, the SST effect has a wide range of potential applications, especially in the fields of ultrashort pulse compression, optical sensing technology and multi-channel light source communication.

The Airy beam was first experimentally generated by Siviloglou et al. in 2007. It possesses the characteristics of self-acceleration, self-healing, and non-diffraction [16]. The Airyprime beam is a beam with new propagation characteristics, which is essentially an extension of the Airy beam [17]. The Airyprime beam is generated by phase modulation of the incident Gaussian beam using a spatial light modulator [18]. In recent years, some researchers have found that Airyprime beams are generated by optical transformation [19]. Specifically, an Airy transformation is performed on the incident beams, which are coherently superimposed from four elegant Hermite-Gaussian modes. Compared with the phase modulation generation method, this method can adjust the intensity, centroid, root mean square width, divergence angle, and propagation factor of the Airyprime beam by changing the width control parameters in the x and y directions, allowing for the flexible selection of the best beam configuration in different application scenarios. The Airyprime beam not only retains the characteristics of the Airy beam but also exhibits an interference enhancement effect [20]. The interference enhancement effect refers to the spontaneous increase in the intensity of the Airyprime beam when it propagates in free space, exceeding the initial light intensity. This is because the Airyprime beam propagates in free space as a coherent superposition of the Airyprime mode and the Airy correlation mode [21], which enhances the energy of the interference part. This characteristic can effectively resist the loss of light intensity caused by atmospheric disturbances [22]. Suppose the Airyprime beam with Cartesian symmetry is extended to a circular Airyprime beam. In that case, it will exhibit stronger self-focusing ability than the traditional circular Airy beam under the same beam parameters, and the focusing intensity will be several times higher [23]. Due to this advantage, the circular Airyprime beam can achieve stronger capture stiffness in particle capture and manipulation in the field of optical tweezers [24]. In the case of introducing optical vortices, the self-focusing ability of the circular Airyprime beam changes significantly with changes in the topological charge and the radial position of the singular point [25,26]. This property enables the beam to generate various special focusing intensity modes and carry orbital angular momentum by adjusting the topological charge and singularity. The Airyprime beam is mathematically isomorphic to the Airyprime pulse. In an experiment, the frequency component is mapped to the time domain by means of dispersion elements (such as fiber or grating) or a pulse modulator so as to transform the characteristics of a spatial Airy beam into the time form of a pulse. The above research demonstrates that the propagation characteristics and potential applications of Airyprime pulses warrant further study on the dynamics of Airyprime pulse propagation.

Researchers have studied the formation and control of SC in single-ZDW [27] and double-ZDW [3] PCFs, as well as the generation of SC in triple-ZDW PCFs with conventional Gaussian pulses [28]. Zhao et al. found that in the PCF with three zero-dispersion wavelengths, the redshifted dispersive wave will form a channel after entering the normal dispersion region, and the energy will be continuously transferred from the initial soliton to the stable tunneling soliton in the next anomalous dispersion region [29]. Rong et al. studied the phenomenon of tunable multi-SST in such PCFs using Airy pulses with small truncation coefficients [3]. Wang Qian et al. discussed the control of SST by symmetric Airy pulses. The results show that the tunneling solitons excited by symmetric Airy pulses not only have higher spectral energy but also can be excited over a shorter propagation distance [30]. However, existing studies have employed limited Raman shifts, typically only once or a few times at most, resulting in relatively limited means for controlling the energy and spectral width of tunneling solitons. Zhang Lifu et al. demonstrated that Airyprime pulses can generate wide-tunable and robust multicolor solitons under Raman effects [31]. Therefore, when an Airyprime pulse is incident onto a zero-dispersion photonic crystal fiber exhibiting Raman effects, it is anticipated that the generated multicolor Raman solitons can be converged into a stable tunneling soliton via the tunneling effect. The author further investigated the influence of the initial Airyprime pulse's truncation coefficient, power, and center position on the tunneling soliton, thereby achieving pulse wavelength conversion. This constitutes a research topic of significant scientific value.

The structure of this paper is as follows: Section 2 introduces the theoretical analysis, covering the generalized nonlinear Schrödinger equation model, the characteristics of triple-zero-dispersion photonic crystal fibers, and the spectral and propagation properties of Airyprime pulses. It also compares the differences between Airyprime pulses and Airy pulses in generating Raman solitons and tunneling solitons. Section 3 presents results and discussion, investigating the effects of cutoff coefficient, peak power, and central wavelength on tunneling soliton formation and evolution while revealing propagation characteristics via X-FROG plots. Section 4 concludes the paper by summarizing key findings and highlighting application prospects in wavelength conversion and supercontinuum control.

2. Theoretical Analysis

2.1. Propagation Model

In the slowly varying amplitude approximation, the light field follows the following generalized nonlinear Schrdinger equation [32]:

$$\begin{aligned}\frac{\partial A(Z,T)}{\partial Z} &= i\sum_{n=2}^{\infty}\frac{i^n}{n!}\frac{L_{D2}}{L_D^{(n)}}\mathrm{sgn}(\beta_n)\frac{\partial^n}{\partial T^n}A(Z,T)-\gamma|A|^2A \\ &-i\gamma\times\left[A(Z,T)\int_{-\infty}^{T}R(T-t')\Big|A(Z,T-t')\,|^2dt'\right]\end{aligned} \tag{1}$$

In Equation (1), the left expression is the propagation term of the pulse, the first term on the right describes the dispersion effect, the second term corresponds to the self-phase modulation, and the third term corresponds to the Raman effect. $A(Z, T)$ represents the evolution of the pulse envelope with the propagation distance Z and the delay time T, β_n is the nth-order dispersion coefficient of the fiber, and γ is the nonlinear coefficient. At the central angular frequency ω_0, the Taylor expansion of the n-order dispersion of the propagation constant β is carried out. $L_D^{(n)} = T_p^n/\beta_n$ represents the nth-order dispersion length, $L_{D2} = \left|L_D^{(2)}\right|$ represents the absolute value of the second-order dispersion length, $Z = z/L_{D2}$ is the normalized propagation distance, and $T = (t - z/v_g)/T_p$ represents the normalized delay time of the pulse propagating at the group velocity. γ is a nonlinear coefficient; $\gamma = 2\pi n_2/\lambda A_{eff}$, and the value in this paper is 0.11. λ is the central wavelength of the pulse. The integral term in Equation (1) describes the energy transfer process caused by stimulated Raman scattering. The nonlinear response function $R(t)$ should include both the electron contribution and the nuclear contribution. Since the electron contribution has near-instantaneous characteristics, the $R(t)$ function can be expressed as:

$$R(t) = (1 - f_R)\delta(t) + f_R\left[\frac{\tau_1^2 + \tau_2^2}{\tau_1\tau_2^2}\right] \times \exp(-t/\tau_2)\sin(t/\tau_1)\Theta(t) \quad (2)$$

The response function contains the contribution of the instantaneous electron effect and the delayed Raman effect, where the causal condition is reflected by the Heavisine function $\Theta(t)$ ($t > 0$ is 1, $t < 0$ is 0); f_R denotes the relative contribution of the delayed Raman response to the nonlinear polarization. The parameters of the response function are $f_R = 0.18$, $\tau_1 = 12.2 f_s/\tau_p$, $\tau_2 = 32 f_s/\tau_p$, and τ_p is the pulse width. The general expression of the nonlinear coefficient γ is:

$$\gamma = \frac{n_2\omega_0}{cA_{eff}} \quad (3)$$

$n_2 \approx 2.2 \times 10^{-20}$ m^2 W^{-1} is the nonlinear refractive index of fused silica, c is the speed of light in vacuum, and A_{eff} is the effective cross-sectional area of the guided wave mode. A_{eff} is defined as:

$$A_{eff} = \frac{\left(\int\int_{-\infty}^{\infty} |F(x,y)|^2 dxdy\right)^2}{\int\int_{-\infty}^{\infty} |F(x,y)|^4 dxdy}. \quad (4)$$

The value of A_{eff} depends on fiber parameters such as core radius and cladding refractive index difference. Nonlinear characteristics can be obtained by appropriately setting the geometric structure (such as spacing and aperture).

2.2. PCF Characteristics

In this paper, a silicon PCF with a length of 0.4 m and three ZDWs is used as the propagation medium. The dispersion coefficient of the fiber is up to ten orders of magnitude: $\beta_2 = -15.8126$ ps^2/km, $\beta_3 = 0.11025$ ps^3/km, $\beta_4 = 1.0382 \times 10^{-3}$ ps^4/km, $\beta_5 = -1.5686 \times 10^{-6}$ ps^5/km, $\beta_6 = 2.4280 \times 10^{-9}$ ps^6/km, $\beta_7 = 2.1260 \times 10^{-10}$ ps^7/km, $\beta_8 = -1.7693 \times 10^{-12}$ ps^8/km, $\beta_9 = 6.4322 \times 10^{-15}$ ps^9/km, $\beta_{10} = 3.1990 \times 10^{-20}$ ps^{10}/km. In addition, the nonlinear coefficient of PCF is $\gamma = 0.35$ w^{-1} m^{-1}.

Figure 1 displays the dispersion curve and relative group delay curve of a PCF featuring three ZDWs. Observing the dispersion curve [Figure 1a] reveals three distinct ZDWs positioned at $\lambda 1 = 771$ nm, $\lambda 2 = 924$ nm, and $\lambda 3 = 974$ nm. These ZDWs partition the PCF's entire wavelength range into four distinct regions: the short-wavelength normal group velocity dispersion (GVD) region (N1: $\lambda \leq \lambda 1$), the short-wavelength anomalous GVD region (A1: $\lambda 1 \leq \lambda \leq \lambda 2$), the long-wavelength normal GVD region (N2: $\lambda 2 \leq \lambda \leq \lambda 3$), and the long-wavelength anomalous GVD region (A2: $\lambda \geq \lambda 3$). Region N1 ($\lambda \leq \lambda 1$): Signals propagate faster at shorter wavelengths, resulting in reduced group delay. Zone A1 ($\lambda 1 \leq \lambda \leq \lambda 2$): Longer wavelength signals propagate faster, and the group delay difference widens. Zone N2 ($\lambda 2 \leq \lambda \leq \lambda 3$): A small region of normal dispersion sandwiched between two anomalous dispersion zones causes a "dip" in group delay within this area. Zone A2 ($\lambda \geq \lambda 3$): Long-wavelength anomalous dispersion dominates, further increasing the group delay difference. The relative group delay variation is evident from the group delay curve (b). As the wavelength increases, the group delay first decreases rapidly, reaching a minimum at 771 nm, and then begins to increase as the wavelength continues to grow. At 924 nm, the group delay reaches a peak. As the wavelength continues to increase, the group delay decreases until 974 nm, after which the group delay curve rises rapidly. Comparing the dispersion curve [Figure 1a] with the group delay curve [Figure 1b] reveals that group delay is the integral of the dispersion coefficient. The zero points (ZDWs) and sign changes in the dispersion curve cause extreme values (peaks/valleys) in the group delay curve.

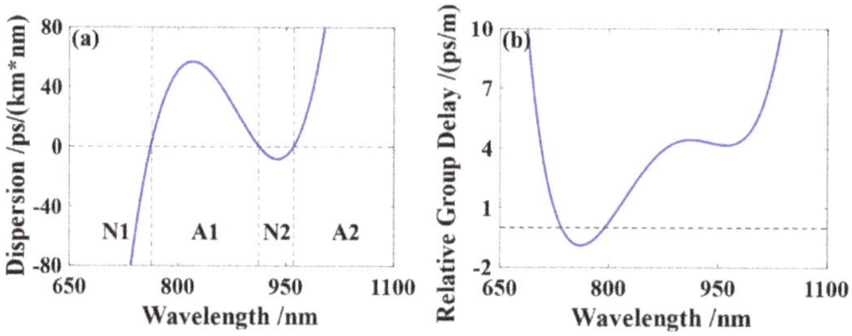

Figure 1. Dispersion curve (**a**) and relative group delay curve (**b**) of PCF with three ZDWs.

2.3. *Initial Pulse Waveform, Spectrum, and X-FROG Diagram Characteristics*

The initial Airyprime pulse of the input can be written in the following form:

$$A(T, Z = 0) = \sqrt{P} Ai'(T) \exp(\alpha T) \tag{5}$$

In the equation, $A(Z, T)$ represents the slow-varying envelope approximation of the complex electric field in the time domain, T denotes the time coordinate in the reference frame moving at group velocity V_g, Z is the propagation distance, P is the initial intensity, $0 < \alpha < 1$ is the truncation coefficient, and $Ai'(T)$ denotes the first derivative of the Airy function. The Airy function is expressed as $Ai(x) = \frac{1}{2\pi}\int_{-\infty}^{\infty} \exp\left[i\left(u^3/3 + xu\right)\right]du$. Performing a Fourier transform on Equation (5) yields the complex amplitude of the Airyprime pulse in the frequency domain:

$$A(\omega, 0) = \sqrt{\alpha^2 + \omega^2} \exp(\alpha^3/3) \exp(-\alpha\omega^2) \times \exp\left[i(\alpha^2\omega - \omega^3/3 + \arctan(\omega/\alpha) + \pi)\right] \tag{6}$$

ω is the frequency of the pulse. The intensity of the initial pulse in the frequency domain is:

$$I = |U(\omega, 0)|^2 = (\alpha^2 + \omega^2) \exp(2\alpha^3/3 - 2\alpha\omega^2) \tag{7}$$

From Equation (7), it is evident that the frequency jointly influences the frequency-domain intensity ω and the cutoff coefficient α. The term a describes the fundamental distribution of spectral intensity, which reaches its minimum at the center frequency $\omega = 0$ and gradually increases as $|\omega|$ grows. The exponential term $\exp(2\alpha^3/3 - 2\alpha\omega^2)$ further modulates the spectral shape, particularly accelerating the intensity decay in the high-frequency region. Overall, the Airyprime pulse frequency distribution exhibits frequency-dependent characteristics: the center frequency corresponds to the point of minimum intensity ($I_{\min} \approx \alpha^2$), while peak intensities in the high and low frequency regions are comparable ($I_{\max} \approx e^{-1}/(2\alpha)$). The peak is located at $\omega_m \approx \pm 1/\sqrt{2\alpha}$, revealing a distribution pattern where intensity increases from low to high frequencies before decreasing.

To comprehensively investigate the evolution of SC across different propagation distances, we performed simultaneous visualization of its temporal and frequency domain evolution. By employing the cross-correlation frequency-resolved optical gating (X-FROG) technique, we recorded the cross-correlation signal between the output pulse at a specific distance and a short-delay tunable reference pulse. This yielded corresponding spectral data, which were used to plot spectral evolution trajectories. The following formula can express the cross-correlation spectrum:

$$S(\tau,\omega) = In|\int_{\infty}^{\infty} A(L,t)A_{ref}(t-\tau)\exp(i\omega t)dt|^2 \tag{8}$$

In the equation, $A(L, t)$ is the amplitude of the initial pulse of length L, $A_{ref}(t-\tau)$ is the amplitude of the reference pulse delayed by a time τ, and $A_{ref}(\tau) = \exp(-\tau^2/\delta^2)$ can be used as an expandable window function. δ is the A_{ref} parameter controlling the Gaussian function.

Figure 2 displays the initial waveform, spectrum, and X-FROG plot of the Airyprime pulse. The initial waveform reveals a pronounced asymmetry [Figure 2a]: The central peak occupies the second position from the right, with the first peak on the right exhibiting rapid decay; multiple oscillatory peaks are distributed to the left of the central peak, their amplitudes diminishing to form a long tail. The initial spectrum reveals that Airyprime pulses exhibit a double-peak structure under different truncation coefficients (α = 0.15, 0.2, 0.25). The two prominent peaks are symmetrically distributed on either side of the center frequency ν_0, approximately at $(\nu - \nu_0)T_0 \approx \pm 0.75$. A low valley exists at the center frequency, forming a notch [Figure 2b]. As the truncation coefficient α increases, the peak intensity of the spectrum decreases and its width narrows. Figure 2c–e shows the X-Frog diagrams of Airyprime pulses at different α values. When α increases from 0.15 to 0.25, the central frequency region where energy is concentrated shrinks, the tailing effect weakens, the energy distribution becomes more focused, and the spectral width simultaneously narrows.

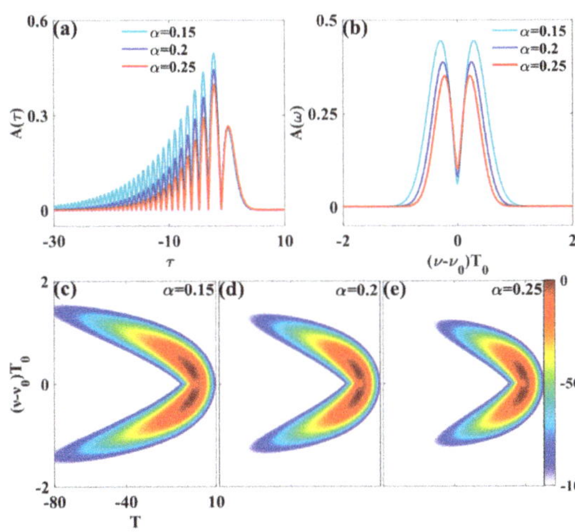

Figure 2. Initial waveform (**a**) and spectrum (**b**) of Airyprime pulses with different truncation coefficients α, as well as the X-Frog images of Airyprime pulses with different truncation coefficients: α = 0.15 (**c**), α = 0.2 (**d**), α = 0.25 (**e**).

3. Results and Discussion

We conducted an in-depth investigation on the temporal and spectral evolution characteristics of Airyprime pulses and Airy pulses in a triple-zero-dispersion medium and performed a comparative analysis of their temporal waveforms and spectral distributions after propagation over 4 m. The results indicate that the Airyprime pulse exhibits significant advantages in generating tunneling solitons. On this basis, we further investigated how adjusting the initial truncation coefficient α, peak power P, and central wavelength λ of the Airyprime pulse can effectively control the energy, pulse width, and central wavelength characteristics of the tunneling solitons generated in the photonic crystal fiber with three zero-dispersion wavelengths.

From the time domain evolution diagram, it can be observed that, under the influence of higher-order dispersion and nonlinear effects, the Airyprime pulse first undergoes temporal compression. It then splits into a series of solitons with different group velocities, simultaneously radiating dispersion wave energy outward [Figure 3a]. Under the effect of Raman self-shift, the split solitons accelerate in a parabolic trajectory, deflecting to the right at a speed proportional to the energy of each peak in the Airyprime pulse. Due to the strong and slowly decaying side lobe energy of the Airyprime pulse, both the central peak and multiple side lobes to its left undergo self-deflection under the Raman effect. In contrast, the Airy pulse exhibits lower side lobe energy, with intensity rapidly decreasing with distance from the central peak, resulting in a significantly smaller number of formed Raman solitons [Figure 3b,c]. From the frequency domain evolution diagram, it can be observed that the Airyprime pulse undergoes brief temporal compression before splitting into multiple static solitons. New frequency components emerge at shorter wavelengths, representing dispersion waves. Subsequently, the split static solitons undergo stimulated Raman scattering, gradually redshifting and linearly shifting toward longer wavelengths. Concurrently, a portion of energy transfers from the redshifted Raman solitons to shorter wavelengths via phase matching under the influence of higher-order dispersion [Figure 3d]. The triple-zero-dispersion wavelength PCF features multiple overlapping normal and anomalous GVD regions within its spectrum. Raman solitons in the anomalous GVD region continuously redshift into the normal GVD region, existing as R-DWs. They form a channel within the normal GVD region connecting two adjacent anomalous GVD regions, continuously transferring energy from the initial soliton to stable tunneling solitons within the subsequent anomalous GVD region. The SST effect can be understood as the process of soliton-induced R-DW generation, involving phase matching between the initial soliton and R-DW, as well as between R-DW and the tunneling soliton. The rapid side lobe attenuation of the Airy pulse resulted in only two Raman self-shift events [Figure 3e]. Since the Airyprime pulse underwent more Raman self-shifts, the resulting tunneling soliton ultimately possessed higher energy [Figure 3d]. The tunneling soliton formed from the Airy pulse exhibits a collapsed structure due to significant deviation in its central position [Figure 3f].

The truncation coefficient α serves as a tuning parameter for the Airyprime pulse waveform, controlling both pulse width and amplitude. A smaller α yields higher peak power and more pronounced Raman effects. Therefore, investigating the influence of the truncation coefficient on Raman effects and spectral tunneling effects is crucial. To this end, numerical simulations were conducted to explore the impact of the truncation coefficient on the generation of spectral tunneling effects in Airyprime pulses. When $\alpha = 0.15$, the Airyprime pulse underwent four Raman self-frequencies. The stable propagation tunneling soliton, formed after traversing the quantum channel via tunneling, exhibited the maximum energy [Figure 4a]. As α increases, the energy of the central peak and side lobes significantly weakens, the Raman effect diminishes, the number of tunneling events decreases, and the spectral width of the tunneling soliton also narrows [Figure 4b,c]. When $\alpha = 0.25$, the center positions of the generated tunneling solitons exhibit a large gap, resulting in a tunneling soliton spectrum with low flatness and a collapsed structure [Figure 4c]. These phenomena can be explained as follows: When α is small, the central peak and right-side lobe of the Airyprime pulse possess substantial energy, promoting Raman self-shifting in the main lobe and part of the side lobes.

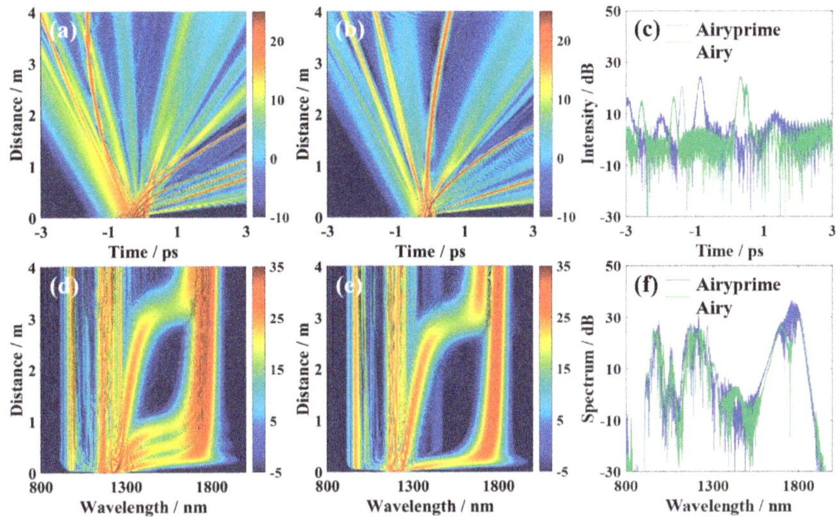

Figure 3. When P_0 = 5000 W, α = 0.175, and λ = 1064 nm, the temporal and spectral evolution of Airyprime and Airy pulses in a PCF with three ZDWs: Airyprime pulse (**a**,**d**); Airy pulse (**b**,**e**). Figure (**c**,**f**) show the corresponding waveforms and spectra of the Airyprime pulse (blue) and Airy pulse (green) at a propagation distance of 4 m.

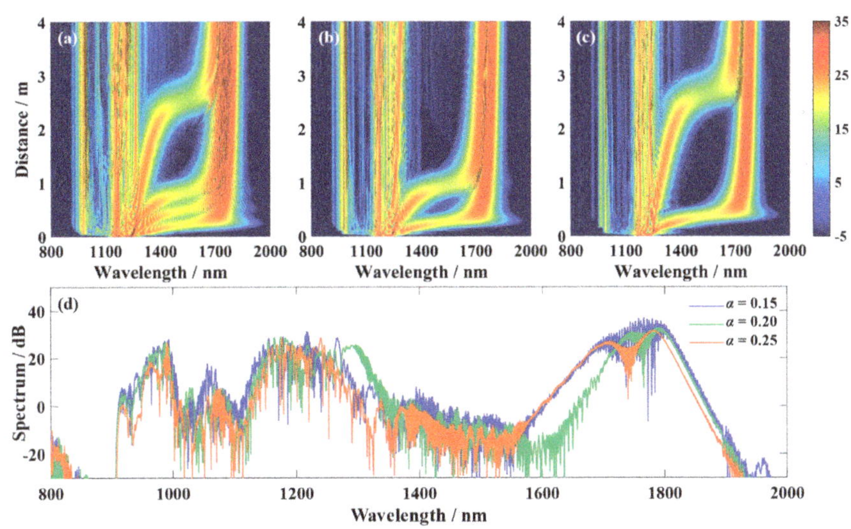

Figure 4. When P = 4000 and λ = 1200 nm, the frequency domain evolution of Airyprime pulses with different truncation coefficients in a PCF with three ZDWs is shown. Figures (**a**–**c**) correspond to α = 0.15, α = 0.2, and α = 0.25, respectively. Figure (**d**) shows the spectra at a propagation distance of 4 m for truncation coefficients of α = 0.15 (blue), α = 0.2 (green), and α = 0.25 (orange).

The influence of power on Raman self-shift is particularly significant in nonlinear optics. Raman self-shift refers to the phenomenon of linear frequency deviation caused by the Raman effect, where light waves interact with excited states within a medium during propagation, resulting in frequency changes. To investigate the influence of peak power P on the SST effect in Airyprime pulses, numerical simulations were conducted in a triple-zero-dispersion PCF with a cutoff coefficient $\alpha = 0.15$ and a center wavelength $\lambda = 1200$ nm. These simulations explored the regulatory role of P in controlling the tunneling soliton in Airyprime pulses. By combining spectral evolution plots with two-dimensional comparison curves, the dynamic evolution of the SST effect was revealed during the P value increment from 3000 W to 5000 W. Redshifted solitons detached from the incident pump pulse and exhibited linear deflection due to the Raman scattering effect. In the time domain, this manifested as changes in pulse trajectory, while in the frequency domain, it appeared as spectral redshift. The Raman self-shift effect causes a significant spectral shift toward longer wavelengths, while the short-wavelength spectrum remains nearly unchanged. At $P = 3000$ W, the Airyprime pulse generates only three Raman solitons, with closely spaced firm peaks in the main spectrum and a relatively low redshift rate [Figure 5a]. As P increases, the Airyprime pulse releases four Raman solitons, with more redshifted peaks separated from the main spectrum [Figure 5b,c]. At $P = 5000$ W, although the number of Raman solitons remains four, their deflection angles significantly increase [Figure 5c]. The spectrum of the Airyprime pulse at a propagation distance of 4 L_D reveals that higher peak power enables stable propagation of Raman solitons tunneling through quantum channels. Although the peak intensity shows little change, the soliton width increases, and its center position undergoes greater redshift [Figure 5d]. Additionally, the intensity of the blue-shifted dispersive wave (B-DW) exhibits linear growth with increasing power. By adjusting the peak power, the number of Raman self-frequency shifts, the deflection angle, and the energy of the soliton after Raman SST can be effectively controlled.

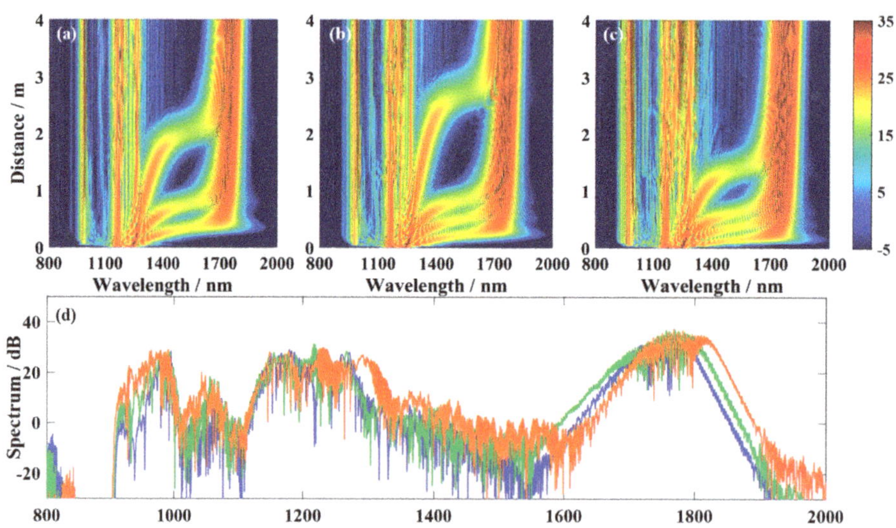

Figure 5. When $\alpha = 0.15$ and $\lambda = 1200$ nm, the frequency domain evolution of Airyprime pulses with different peak powers P in a PCF with three ZDWs. Among them, (**a**) $P = 3000$ W, (**b**) $P = 4000$ W, (**c**) $P = 5000$ W. Figure (**d**) shows the corresponding spectra at a propagation distance of 4 m for peak powers of $P = 3000$ (blue), $P = 4000$ (green), and $P = 5000$ (orange).

The central wavelength has an important influence on the generation of SC, especially in the optical fiber process. SC refers to the phenomenon where the spectrum expands to a broader frequency band through the interaction between an intense light pulse and the medium. SC usually depends on the nonlinear propagation characteristics of the pulse, where the central wavelength is a key factor that directly affects the width, intensity distribution, and generation process of SC. Figure 6 illustrates the generation of tunneling solitons in PCF with zero dispersion at different central wavelengths λ for Airyprime pulses. At λ = 1000 nm, the spectrum broadens toward both the short-wavelength (800 nm) and long-wavelength (1300 nm) regions [Figure 6a]. As the central wavelength increases, although the degree of spectral linear shift due to Raman scattering effects changes little, higher-order soliton generation becomes pronounced [Figure 6b,c]. This is accompanied by highly efficient redshifted dispersion waves, extending the tunneling soliton spectrum to 1800 nm on the long-wavelength side. Concurrently, the B-DW broadening diminishes to 900 nm. The spectrum of the Airyprime pulse at a propagation distance of $4L_D$ reveals. At λ = 1000 nm, the energy enhancement of the tunneling soliton is limited due to the dispersion-induced redshift effect, resulting in the narrowest width. At λ = 1200 nm, the tunneling intensity increases, leading to a significant expansion in both the width and central wavelength position of the tunneling soliton [Figure 6d]. By adjusting the central wavelength of the pulse, the formation and evolution of tunneling solitons can be effectively controlled, thereby regulating the broadening range and energy distribution characteristics of the tunneling soliton spectrum.

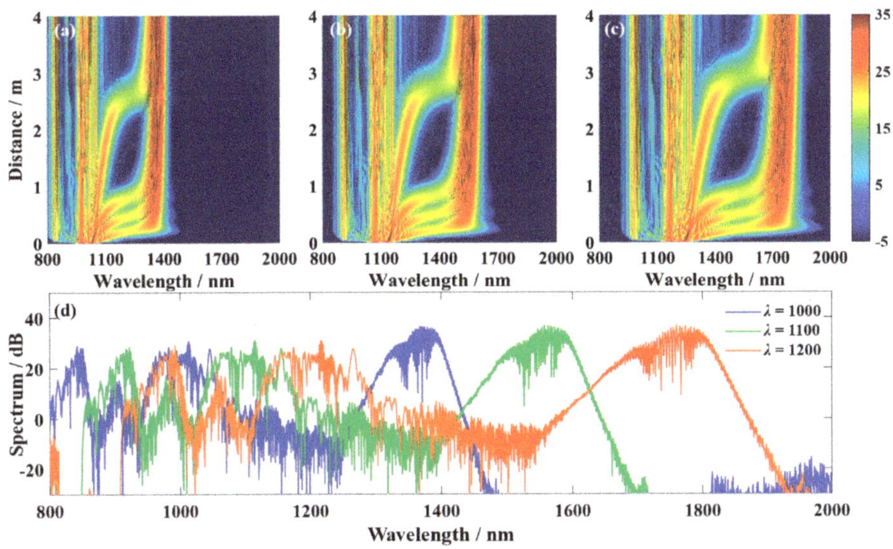

Figure 6. When α = 0.15 and P = 4000, the frequency domain evolution of Airyprime pulses with different central wavelengths λ in a PCF with three ZDWs. Among them, (**a**) λ = 1000, (**b**) λ = 1100, (**c**) λ = 1200. Figure (**d**) shows the corresponding spectra at a propagation distance of 4 m for central wavelengths of λ = 1000 (blue), λ = 1100 (green), and λ = 1200 (orange).

Figure 7 illustrates the influence of different parameters on the width of tunneling solitons in a zero-dispersion wavelength PCF. It can be observed that as the cutoff coefficient increases, the soliton width first decreases and then increases, exhibiting a distinct U-shaped trend. When the cutoff coefficient approaches approximately 0.2, the width of the tunneling soliton reaches its minimum value [Figure 7a]. When the cutoff coefficient exceeds 0.2, the continued increase in soliton width stems from significant positional deviation of the soliton center caused by multiple tunneling effects. Although the soliton width increases at this point, its flatness decreases markedly. Additionally, as the power increases, the tunneling soliton width continues to grow, exhibiting an overall trend of near-linear increase. This indicates that higher power leads to broadening of the tunneling soliton [Figure 7b]. Finally, the tunneling soliton width increases significantly with increasing central wavelength, exhibiting a clear monotonic upward trend. This demonstrates that longer central wavelengths correspond to broader tunneling soliton spectral widths [Figure 7c]. Overall, the cutoff coefficient, input power, and central wavelength all significantly influence the width of the tunneling soliton: a smaller cutoff coefficient results in a broader soliton with higher flatness; input power is positively correlated with soliton width; and a longer central wavelength leads to a broader soliton.

To gain deeper insight into the formation process of the SST effect during pulse propagation, Figure 7 presents the X-FROG traces of output pulses for six different propagation distances when an Airyprime pulse with parameters $\lambda = 1200$, $P = 4000$, and $\alpha = 0.15$ propagates in a PCF featuring three zero-dispersion points. It can be seen from Figure 8a that the Airyprime pulse signal has a highly narrow pulse width in the short-wavelength anomalous GVD region at the input end of the PCF. When the pulse is transmitted to the $Z = 0.2$ position, it can be observed that the SPM-dominated spectrum broadening basically ends, and the higher-order solitons have begun to split. Because of the influence of nonlinear effects, the spectrum broadening is asymmetric. When the pulse signal is transmitted to 0.4 m, as shown in Figure 8c, due to the high-order nonlinear effect and dispersion, the high-order solitons begin to split and emit the first fundamental soliton, accompanied by B-DW and R-DW in N1 and N2, respectively. At this time, the R-DW in N2 forms an energy channel. From Figure 8d, it can be seen that under the action of SSFS, the fundamental solitons are continuously redshifted, and the number of quantum channels increases. In the process of the soliton self-frequency shift effect, as shown in Figure 8e, the energy of the initial soliton in A1 is continuously transferred to the tunneling soliton in A2, and the number of quantum channels reaches the maximum. As the propagation distance continues to increase, the number of quantum channels remains unchanged. By reasonably adjusting the propagation distance, we can effectively control the generation time and spectral width of tunneling solitons.

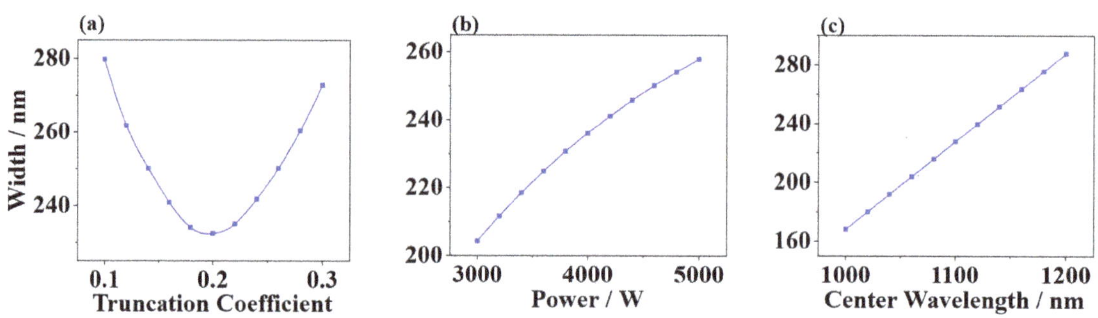

Figure 7. The influence of different (**a**) truncation coefficients, (**b**) initial chirps, and (**c**) distribution factors on the width of tunneling solitons in photonic crystal fibers with zero-dispersion wavelength.

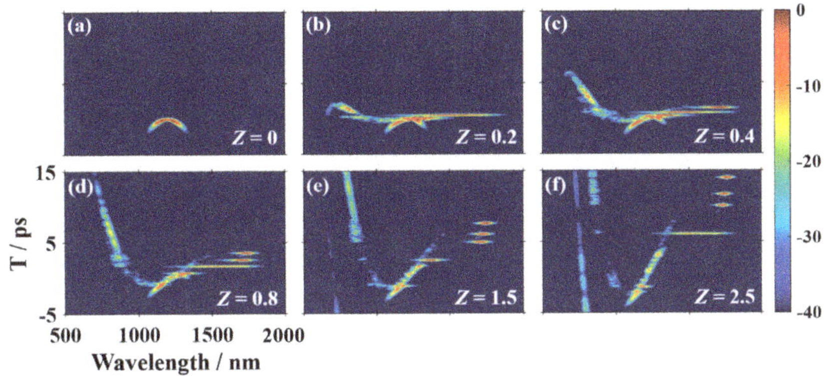

Figure 8. The frog diagrams of Airyprime at different propagation distances, with wavelength = 1200; $P = 4000$; $\alpha = 0.15$.

4. Conclusions

In this paper, the generation of multicolor Raman solitons and tunneling solitons based on Airyprime pulses and their propagation characteristics in a triple-zero-dispersion wavelength PCF are studied. By adjusting the parameters of the Airyprime pulse, such as the truncation coefficient, input power, and center wavelength, we can discuss their effects on the SST effect. The results show that the Airyprime pulse has strong nonlinear effects and self-accelerating characteristics. It can generate multiple Raman solitons in the fiber and form stable tunneling solitons through the SST effect. This process is not only closely related to the Raman self-frequency shift but also significantly affected by the dispersion characteristics and nonlinear effects. By adjusting the truncation coefficient of the Airyprime pulse, the energy, width, and central wavelength characteristics of tunneling solitons can be effectively controlled. It is found that the smaller the truncation coefficient, the stronger the Raman effect and the greater the energy of the tunneling soliton; the width also increases. As the truncation coefficient increases, the number of Raman self-frequency shifts decreases, the final tunneling soliton energy is small, and the spectral width is narrow. In addition, the input power and central wavelength also have an essential influence on the characteristics of tunneling solitons. As the power increases, the width of the tunneling soliton significantly increases, and the energy distribution becomes more concentrated. The change in the central wavelength mainly affects the redshift degree and width of the tunneling soliton. The longer central wavelength leads to an increase in the width of the tunneling soliton, causing the spectrum to expand to longer wavelengths. These results provide a theoretical basis for more accurate wavelength conversion by adjusting the initial pulse parameters. It is also shown that in PCFs with multiple ZDWs, the SST effect can effectively realize applications such as ultrashort pulse compression, optical sensing technology, and multi-channel light source communication. Through in-depth analysis of the propagation of Airyprime pulses in PCF, this paper provides new ideas and methods for PCF-based SC generation and wavelength conversion technology. In summary, Airyprime pulses show great potential in Raman soliton generation, tunneling effect, and wavelength control, which will be used in future ultrafast optical devices and optical communication technology. Future research can be carried out in the following directions: Further studies can be conducted on the influence of different PCF structures—such as asymmetric cladding designs or adjustable air-hole arrangements—on the propagation characteristics of Airyprime pulses in order to achieve more precise control over the conditions for tunneling soliton generation. In addition, the effects of other types of abrupt pulses, such as Pearce pulses, swallowtail pulses, and butterfly pulses, on the formation of tunneling solitons can be explored to broaden the understanding of nonlinear pulse dynamics. Finally, future work can combine experimental investigations with numerical simulations to verify the theoretical results and provide stronger support for the proposed model.

Author Contributions: Conceptualization, H.W. and B.W.; methodology, H.W.; validation, H.W.; data curation, B.W.; writing—original draft preparation, M.D.; writing—review and editing, M.D.; visualization, M.D.; supervision, H.W.; funding acquisition, M.D. All authors have read and agreed to the published version of the manuscript.

Funding: This research is funded by the Loudi City Science and Technology Planning Project under grant number ld2017016.

Data Availability Statement: The original contributions presented in this study are included in the article. Further inquiries can be directed to the corresponding author(s).

Conflicts of Interest: The authors declare no conflicts of interest.

References

1. Petersen, C.R.; Engelsholm, R.D.; Markos, C.; Brilland, L.; Caillaud, C.; Trolès, J.; Bang, O. Increased mid-infrared supercontinuum bandwidth and average power by tapering large-mode-area chalcogenide photonic crystal fibers. *Opt. Express* **2017**, *25*, 15336–15348. [CrossRef]
2. Guo, H.R.; Wang, S.F.; Zeng, X.L.; Bache, M. Understanding soliton spectral tunneling as a spectral coupling effect. *IEEE Photonics Technol. Lett.* **2013**, *25*, 1928–1931. [CrossRef]
3. Rong, J.; Yang, H.; Ran, C.; Xu, Q.; Liu, X. Controllable multi-soliton spectral tunneling by Airy pulses in photonic crystal fiber. *Opt. Commun.* **2024**, *569*, 130731. [CrossRef]
4. Poletti, F.; Horak, P.; Richardson, D.J. Soliton spectral tunneling in dispersion-controlled holey fibers. *IEEE Photonics Technol. Lett.* **2008**, *20*, 1414–1416. [CrossRef]
5. Yang, H.; Xiao, G.Y.; Zhao, S.L.; Tang, Z.; Li, T.; Luo, Y.; Tian, X. Impact of the self-steepening effect on soliton spectral tunneling in PCF with three zero dispersion wavelengths. *Chin. Opt. Lett.* **2018**, *16*, 070601. [CrossRef]
6. Wang, S.F.; Hu, J.G.; Guo, H.R.; Zeng, X. Optical Cherenkov radiation in an As_2S_3 slot waveguide with four zero-dispersion wavelengths. *Opt. Express* **2013**, *21*, 3067–3072. [CrossRef]
7. González-Baquedano, N.; Torres-Gómez, I.; Arzate, N.; Ferrando, A.; Ceballos-Herrera, D.E. Pulse quality analysis on soliton pulse compression and soliton self-frequency shift in a hollow-core photonic bandgap fiber. *Opt. Express* **2013**, *21*, 9132–9143. [CrossRef]
8. Braud, F.; Conforti, M.; Cassez, A.; Kudlinski, A. Solitonization of a dispersive wave. *Opt. Lett.* **2016**, *41*, 1412–1415. [CrossRef]
9. Marest, T.; Braud, F.; Conforti, M.; Wabnitz, S.; Mussot, A.; Kudlinski, A. Longitudinal soliton tunneling in optical fiber. *Opt. Lett.* **2017**, *42*, 2350–2353. [CrossRef]
10. Luo, Y.P.; Yang, H.; Zhao, S.L.; Lv, J.; Hu, H. Impact of potential barriers on soliton spectral tunneling in photonic crystal fiber. *Opt. Commun.* **2020**, *454*, 124330. [CrossRef]
11. Zhao, S.L.; Yang, H.; Zhao, Y.L.; Xiao, Y. Supercontinuum manipulation based on the influence of chirp on soliton spectral tunneling. *Chin. Phys. B* **2018**, *27*, 114219. [CrossRef]
12. Dudley, J.M.; Genty, G.; Coen, S. Supercontinuum generation in photonic crystal fiber. *Rev. Mod. Phys.* **2006**, *78*, 1135–1184. [CrossRef]
13. Stark, S.P.; Biancalana, F.; Podlipensky, A.; Russell, P.S.J. Nonlinear wavelength conversion in photonic crystal fibers with three zero-dispersion points. *Phys. Rev. A* **2011**, *83*, 023808. [CrossRef]
14. Yu, X.Y.; Wen, B.; Deng, Y.B.; Gao, C. Supercontinuum Generation from Airy-Gaussian Pulses in Photonic Crystal Fiber with Three Zero-Dispersion Points. *Photonics* **2023**, *10*, 1061. [CrossRef]
15. Huang, Y.; Yang, H.; Zhao, S.; Mao, Y.; Chen, S. Design of photonic crystal fibers with flat dispersion and three zero dispersion wavelengths for coherent supercontinuum generation in both normal and anomalous regions. *Results Phys.* **2021**, *23*, 104033. [CrossRef]
16. Siviloglou, G.A.; Broky, J.; Dogariu, A.; Christodoulides, D.N. Observation of Accelerating Airy Beams. *Phys. Rev. Lett.* **2007**, *99*, 213901. [CrossRef]
17. Zhou, G.Q.; Chen, R.P.; Ru, G.Y. Airyprime beams and their propagation characteristics. *Laser Phys. Lett.* **2015**, *12*, 025003. [CrossRef]
18. Zheng, X.; Yang, Y.; Liu, Y.; Lin, X.; Liang, Z.; Liu, J.; Deng, D. Adjustable focusing property of circular Airyprime beam through Fourier space modulation. *Optics Lett.* **2024**, *49*, 4393–4396. [CrossRef]

19. Zang, X.; Dan, W.; Wang, F.; Zhou, Y.; Xu, Y.; Zhou, G. Generation of finite energy Airyprime beams by Airy transformation. *Opt. Express* **2022**, *30*, 24948–24967. [CrossRef]
20. Dan, W.; Zang, X.; Wang, F.; Zhou, Y.; Xu, Y.; Chen, R.; Zhou, G. Interference enhancement effect in a single Airyprime beam propagating in free space. *Opt. Express* **2022**, *30*, 32704–32721. [CrossRef]
21. Dan, W.; Zang, X.; Wang, F.; Chu, X.; Ni, Y.; Cai, Y.; Zhou, G. Investigation of the effect of chirped factors on the interference enhancement effect of an Airyprime beam propagating in free space. *Opt. Express* **2023**, *31*, 10820–10839. [CrossRef]
22. Zhao, H.; Huang, X.; Bai, Y.; Fu, X. Propagation properties of a finite energy Airyprime beam in atmospheric turbulence. *J. Opt. Soc. Am. A* **2025**, *42*, 862–869. [CrossRef]
23. Zang, X.; Dan, W.; Zhou, Y.; Lv, H.; Wang, F.; Cai, Y.; Zhou, G. Abruptly autofocusing of generalized circular Airy derivative beams. *Opt. Express* **2022**, *30*, 3804–3819. [CrossRef]
24. Dan, W.-S.; He, J.; Zang, X.; Wang, F.; Zhou, Y.-M.; Chu, X.-X.; Zhou, G.-Q. Abrupt autofocusing performance of a circular Airyprime beam with vortex pairs. *Results Phys.* **2023**, *47*, 106377. [CrossRef]
25. Zhao, S.; Huang, X.; Bai, Y.; Fu, X. Abruptly autofocusing properties and radiation forces of an Airy derivative beam array with an optical vortex. *Chaos Solitons Fractals* **2024**, *187*, 115480. [CrossRef]
26. Li, Y.; Peng, J.; Tan, C.; Tang, P. Numerical investigation on the dynamic propagation of finite-energy Airy pulses in photonic crystal fiber. *J. Mod. Opt.* **2020**, *67*, 445–453. [CrossRef]
27. Peng, H.; Li, Y.; Peng, J.; Wen, B.; Deng, Y.; Tang, P. Evolution of Airy-Gaussian pulses in photonic crystal fiber with two zero-dispersion wavelengths. *Opt. Int. J. Light Electron Opt.* **2021**, *250*, 168324. [CrossRef]
28. Zang, X.; Wang, F.; Dan, W.; Zhou, Y.; Zhou, G. Propagation dynamics of abruptly autofocusing circular Airyprime beam with an optical vortex. *Opt. Laser Technol.* **2022**, *155*, 108398. [CrossRef]
29. Zhao, S.; Yang, H.; Chen, N.; Fu, X.; Zhao, C. Soliton Trapping of Dispersive Waves in Photonic Crystal Fiber with Three Zero-Dispersive Wavelengths. *IEEE Photonics J.* **2015**, *7*, 7102709.
30. Wang, Q.; Xu, C.; Gao, Y.; Wang, Z.; Fan, D.; Zhang, L. Controlling soliton spectral tunneling via symmetric Airy pulses at the advanced stage of supercontinuum generation. *Opt. Commun.* **2025**, *596*, 132480. [CrossRef]
31. Zhang, L.; Yang, X.; Huang, Q.; Wang, Z.; Gao, Y.; Fan, D. Widely Tunable and Robust Multicolor Raman Solitons in Nonlinear Fiber Through Pump Modulation: A Numerical Study. *J. Light. Technol.* **2025**, *43*, 5007–5012. [CrossRef]
32. Shi, X.H.; Tan, C.; Bai, Y.F.; Wang, R.; Zhang, Z.; Fu, X. Dressed dynamics of two time-reversed shapes of Airy pulses in a relaxing nonlinear medium. *J. Opt. Soc. Am. B* **2025**, *32*, 1816–1823. [CrossRef]

crystals

MDPI

Article

Line-Defect Phononic Crystal Structure for Directional Enhancement Detection of Weak Acoustic Signals

Shijie Zhang [1], Jinling Mu [1], Jiawei Xiao [2,3,*] and Huiqiang Xu [4]

1. School of Artificial Intelligence, Luoyang Normal University, Luoyang 471934, China; zhangshijie@lynu.edu.cn (S.Z.); 221114015@lynu.edu.cn (J.M.)
2. Hubei Key Laboratory of Hydroelectric Machinery Design & Maintenance, China Three Gorges University, Yichang 443002, China
3. College of Mechanical and Power Engineering, China Three Gorges University, Yichang 443002, China
4. China Xuchang Xuji Wind Power Technology Co., Ltd., Xuchang 461000, China; xuhuiqiang.xj@cee.com.cn

* Correspondence: xiaojiawei@ctgu.edu.cn

Abstract: Effective detection of acoustic signals plays a crucial role in numerous fields, including industrial equipment fault prediction and environmental monitoring. Acoustic sensing technology, owing to its substantial information carrying capacity and non-contact measurement advantages, has garnered widespread attention in relevant applications. However, the effective detection of weak target acoustic signals amidst strong noise interference remains a critical challenge in this field. The core bottleneck lies in the difficulty of traditional detection methods to simultaneously achieve both high sensitivity and high directionality. To address this limitation, this work proposes a line-defect phononic crystal (PnC) structure that enables directional enhancement and detection of weak target signals under intense spatial noise interference by coupling defect state localization characteristics with anisotropy mechanisms. Through theoretical derivation and finite element numerical simulation, the directional enhancement properties of this structure were systematically validated. Furthermore, numerical simulations were conducted to validate the detection of weak harmonic signals and weak bearing fault signals under strong spatial noise interference. The results demonstrate that this line-defect phononic crystal (PnC) structure exhibits high feasibility and outstanding performance in detecting weak acoustic signals. This work provides novel insights for developing new acoustic detection methods combining high sensitivity with high directivity, showcasing unique advantages and broad application prospects in acoustic signal sensing, enhancement, and localization.

Keywords: phononic crystals; defect state; directional enhancement; weak acoustic signal detection

1. Introduction

Acoustic signals are energy signals propagating through media such as gases and liquids, carrying rich physical information. Acoustic directional detection plays a crucial role in numerous fields, with typical applications including microphone design [1–3], acoustic communication [4–6] and industrial equipment health monitoring [7–9]. The core method for traditional acoustic directional detection is microphone array technology, but its detection accuracy is severely constrained by the performance of signal processing algorithms and the physical dimensions of the array [10–12]. Additionally, biomimetic microphones inspired by insects in nature offer novel approaches to acoustic directional detection. Examples include bionic directional perception microphones developed by mimicking the sensory characteristics of micro-insects such as Fly *Ormia ochracea* [13] and *Achroia grisella* [14]. However, while these technologies have opened new avenues for acoustic directional detection, spatial noise from multiple directions easily overlaps with target acoustic signals in high-noise environments, causing detection accuracy to plummet or even fail. Therefore, developing novel acoustic direction-finding methods to overcome the technical bottleneck of detecting weak acoustic signals holds significant theoretical value and engineering importance.

Over the past two decades, phononic crystals (PnCs) have emerged as artificial structures capable of manipulating acoustic wave propagation, garnering significant academic and engineering interest through their unique physical properties. Consequently, PnCs demonstrate broad application prospects in acoustic wave control, including but not limited to acoustic superlensing [15–17], acoustic cloaking [18–20], acoustic focusing [21–24], and acoustic diodes and acoustic noise reduction [25–28]. These characteristics establish a foundation for revolutionizing directional acoustic detection, enabling novel methodologies with high sensitivity and directivity. For instance, Miniaci et al. [29] proposed a hypersensitive PnC device that demonstrates filtering and spatial focusing functionalities within specific target frequency ranges. When integrated with time reversal techniques, this device allows for the detection and localization of nonlinear elastic sources, including cracks and delaminations. Ma et al. [30] designed a PnC structure with near-zero refractive index, achieving effective detection of weak Gaussian distributed sinusoidal pulse signals submerged within 30° spatial noise. Xiao et al. [31] created a phononic crystal resonator (PnCR) with directional enhancement capability, which successfully detected gear fault signals under 104.4 dB sound pressure level noise interference. PnCs with defect patterns enable the localization and amplification of acoustic energy, demonstrating significant potential in numerous novel applications such as acoustic sensing [32,33], tunable actuation [34,35], and acoustic energy harvesting [36,37]. Recent studies [38–41] report novel defect-mode PnCs with integrated acoustic enhancement and directional sensing functions through coupled-resonator configurations. However, resonator-coupled designs inevitably increase structural complexity and manufacturing costs. Approaches leveraging intrinsic structural properties of defect-mode PnCs for directional enhancement remain underexplored.

Compared to coupled resonator PnCs, anisotropic PnCs offer the advantages of simpler structural design and lower manufacturing costs [42,43]. Inspired by anisotropic PnCs, this paper proposes a line-defect PnC structure that combines high sensitivity with high directivity. It aims to overcome the bottleneck in detecting weak acoustic signals in strong noise environments by achieving directional enhancement detection. To achieve this goal, a unit cell structure featuring an inner rectangular cavity intersecting with a straight channel is designed. This unit cell is then periodically arranged in two dimensions to construct a line-defect supercell. Through deep analysis of the structural dispersion relations and acoustic transmission properties, the physical mechanism enabling acoustic enhancement and directional sensing in the proposed line-defect PnC is elucidated. Furthermore, numerical simulation results demonstrate that the designed line-defect PnC exhibits superior directional enhancement detection performance for both weak harmonic signals and weak bearing fault signals under strong complex spatial noise interference.

2. Structural Design and Theoretical Calculations

2.1. Structural Design

To achieve directional enhancement of incident acoustic signals, this paper proposes a periodically distributed anisotropic line-defect PnC structure, as shown in Figure 1a. Based on the design principles of anisotropic PnCs [42,43], the unit cell was designed with a penetrating rectangular cavity and straight channels. Meanwhile, in consideration of structural fabrication and application constraints, the key geometric parameters were designed as follows: lattice constant a = 60 mm, inner rectangular cavity length b = 44 mm and straight channel width c = 10 mm, as shown in Figure 1b. The white region corresponds to air, characterized by a mass density of $\rho_{\text{air}} = 1.21\ \text{kg/m}^3$ and a sound velocity of c_{air} = 343 m/s. Considering structural practicality and manufacturability, photosensitive resin (blue regions in Figure 1b) was selected with material properties: with the mass density $\rho_1 = 1050\ \text{kg/m}^3$, sound speed c_1 = 2540 m/s, Young's modulus E1 = 5.08 GPa and Poisson's ratio μ = 0.35, respectively. By periodically arranging structures in the x and y directions, one may obtain an ideal two-dimensional square lattice perfect PnC. Figure 1c illustrates a 4 × 1 perfect supercell. As depicted in Figure 1d, removing the two central columns of the perfect supercell (blue shaded region in Figure 1c) induces a line-defect cavity (orange shaded region in Figure 1d), thereby generating Anderson localization effects for acoustic energy [44].

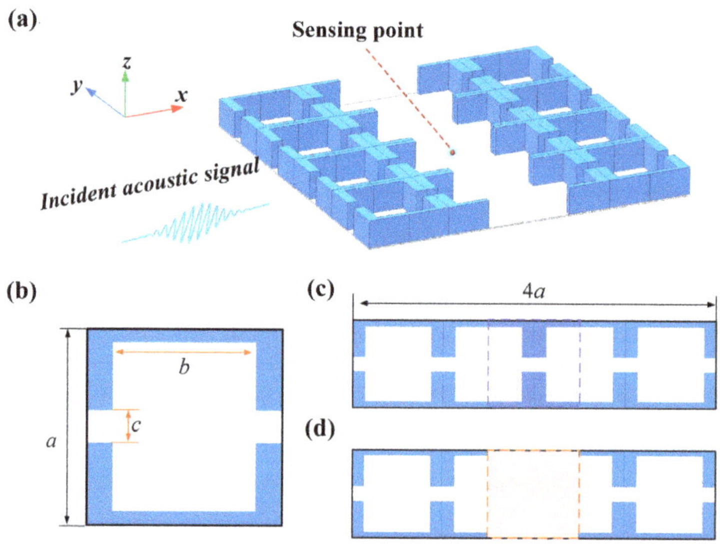

Figure 1. (**a**) Three-dimensional view of the proposed directional enhanced line-defect PnC; (**b**) Unit cell; (**c**) 4 × 1 supercell of perfect PnC and (**d**) line-defect PnC. The blue shaded areas indicate sections to be removed, and the orange shaded areas represent line-defect cavities.

2.2. Theoretical Calculation of Dispersion Relations

Dispersion relation analysis serves as a fundamental approach for investigating wave-manipulation characteristics in PnCs. The finite element method (FEM) is commonly employed to investigate acoustic wave propagation in PnCs owing to its computational efficiency, superior convergence, and ability to accommodate arbitrarily complex structures [45–47]. In this study, the dispersion relations of the proposed structure were calculated using FEM combined with supercell analysis. The propagation of sound waves in periodic structures is governed by the Helmholtz equation [48]:

$$\nabla \cdot \left[\frac{1}{\rho_{air}(\mathbf{r})} \nabla P(\mathbf{r}) \right] + \frac{\omega^2}{\rho_{air}(\mathbf{r}) c_{air}(\mathbf{r})^2} P(\mathbf{r}) = 0 \tag{1}$$

where $P(\mathbf{r})$ represents the sound pressure field, $\mathbf{r}$ = {x, y} denotes the position vector, and ω stands for the angular frequency. In line with Bloch's theory, $P(\mathbf{r})$ within a periodic structure can be expressed as:

$$P(\mathbf{r}) = e^{i\mathbf{k}\cdot\mathbf{r}-\omega t} \sum\nolimits_{\mathbf{G}_m} \boldsymbol{\Phi}_{\mathbf{k}}(\mathbf{G}_m) \cdot e^{i(\mathbf{G}_m \cdot \mathbf{r})} \tag{2}$$

where $\mathbf{k} = \{k_x, k_y\}$ denotes the Bloch wave vector within the first irreducible Brillouin zone (IBZ), $\mathbf{\Phi_k}$ is the scalar potential, and $\mathbf{G}_m$ stands for the reciprocal lattice vector. The scalar potential $\mathbf{\Phi}(\mathbf{r}, t)$ can be used to define this equation, which is expressed as follows [31]:

$$\frac{1}{\rho_{air}c_{air}^2}\frac{\partial^2\Phi}{\partial t^2} = \nabla \cdot \left(\frac{1}{\rho_{air}}\nabla\Phi\right) \tag{3}$$

Assuming the initial harmonic sound pressure is $P = P_0e^{i\omega t}$, Equation (3) may be succinctly rewritten [49]:

$$\frac{\omega^2 P_0}{\rho_{air}c_{air}^2} + \nabla \cdot \left(-\frac{1}{\rho_{air}}\nabla P\right) = 0 \tag{4}$$

As a result, the complete band structure of the structure can be obtained by solving the dispersion relation between the wave vector **k** and angular frequency ω in Equation (4). The finite element analysis software COMSOL Multiphysics 6.2 was utilized to compute the band structures of the perfect PnC supercell and the line-defect PnC supercell within the first IBZ, and these results are illustrated in Figure 2a,b. In the perfect PnC, an initial band gap exists within the frequency range of 1360 Hz to 3360 Hz (grey region in Figure 2a), indicating that sound waves within this range cannot propagate through the PnC structure. Introducing defects alters the original band gap profile, resulting in new band gaps and generating a defect band near 3324 Hz. This physical phenomenon enables the confinement of evanescent waves within the defect core and adjacent regions, achieving acoustic energy localization as depicted in Figure 2c. Typically, when using defect PnC for acoustic signal detection, the sound pressure signal at the center point is primarily measured to obtain the local sound pressure within the defect PnC. In actual experiments, micro-electro-mechanical system (MEMS) acoustic sensing units can be employed for measurement. This unusual physical property can be harnessed to boost the detection of acoustic signals, in turn enhancing the capabilities of sensing and manipulating sound waves.

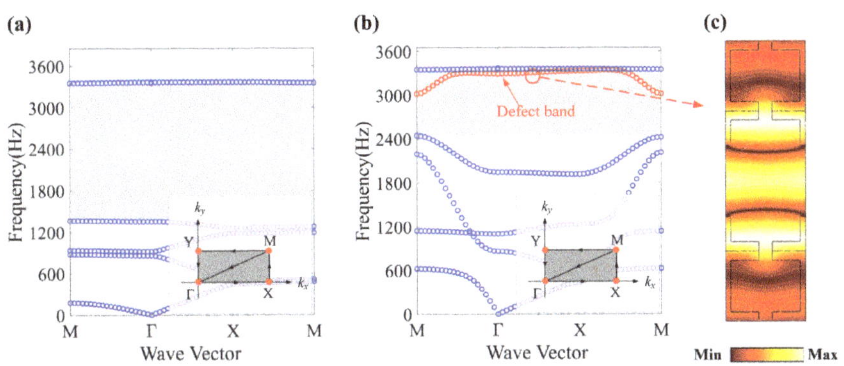

Figure 2. The band structure of (**a**) the perfect PnC and (**b**) the line-defect PnC. (**c**) Absolute sound pressure distribution of the defective mode propagation solution. The inset shows the first irreducible Brillouin zone, and the red circle indicates the defect band.

2.3. Theoretical Calculation of Directional Enhancement Transmission

To accurately characterize the acoustic transmission properties of the designed line-defect PnC, the acoustic analysis module of COMSOL Multiphysics was used to establish a computational analysis model for the structure, and theoretical numerical calculations were performed on the transmission of the line-defect PnC structure. It should be noted that considering the application scenario of the designed line-defect PnC, where sound waves propagate along the *x* and *y* planes (parallel to the structure surface), a Two-dimensional

(2D) simulation model was adopted in this study, as illustrated in Figure 3. The 2D model not only enables analysis of the propagation characteristics of the PnC but also significantly reduces computational complexity. This approach has been widely applied in PnC transmission calculations [31,33,39]. In the acoustic transmission calculation model, the incident plane wave propagates from left to right, the amplitude P_0 is set to 1 Pa, and the incident angle α is 0°. Given that the material impedance of this structure is significantly higher than that of air, its boundary is defined as a hard acoustic boundary. Additionally, the top and bottom boundaries are configured as rigid boundaries to generate ideal plane waves. Meanwhile, the right boundary is designated as a perfectly matched layer boundary to absorb unwanted acoustic reflections.

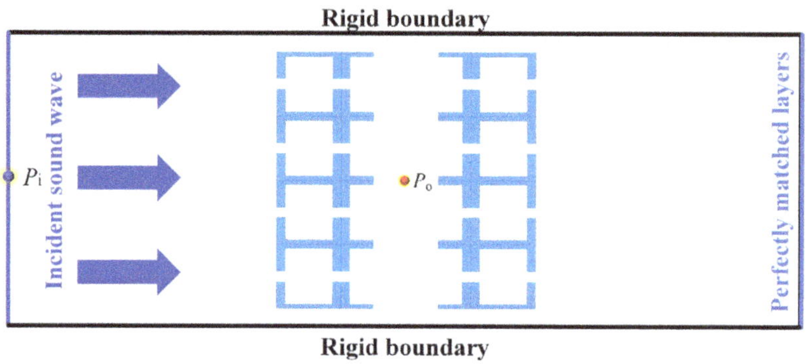

Figure 3. 2D schematic diagram for acoustic transmission analysis of design line-defect PnC.

To calculate the acoustic enhancement effect of the line-defect PnC, the midpoint of the incident boundary is designated as the acoustic incident point P_i, and the midpoint position of the line-defect PnC is set as the measurement point sound pressure P_o. Therefore, the local measurement point sound pressure transmission power T_p of the designed line-defect PnC is denoted as [50]:

$$T_p = 10\lg\frac{P_{out}}{P_{in}} \tag{5}$$

where incident acoustic power P_{in} and measured acoustic power P_{out} can be expressed as:

$$P_{in} = \int_{\partial\Omega_1} \frac{|p_i|^2}{2\rho_0 c_0} dS, P_{out} = \int_{\partial\Omega_2} \frac{|p_o|^2}{2\rho_0 c_0} dS \tag{6}$$

where p_i denotes the sound pressure at the incident point P_i, and p_o denotes the sound pressure at the measurement point P_o, $\partial\Omega_1$ and$\partial\Omega_2$ denote the integral at the incident point and measurement point, respectively. S represents the connected acoustic boundary. Furthermore, the sound pressure gain PG of the structure is expressed as:

$$PG = \frac{P_{out}}{P_{in}} \tag{7}$$

It should be noted that the designed line-defect PnC structure has narrow channels, and acoustic viscous effects must be taken into account to obtain more accurate numerical solutions for acoustic propagation. We employed the boundary layer impedance method [51,52] to simulate the dissipative effects arising from acoustic viscous forces at all boundaries of the line-defect PnC. To obtain the optimal gain structure, we calculated the sound pressure transmission power spectrum and pressure gain spectrum of the line-defect PnC for different straight channel widths c, with results shown in Figure 4a,b. The results indicate a maximum gain resonance peak near 3324 Hz, corresponding to the band frequency of the structural line defect. It has a transmission power exceeding 24.5 dB and a sound pressure gain exceeding 176.9 times. Therefore, considering the optimal balance between sound pressure gain and structural dimensions, we selected a straight channel width of c = 10 mm.

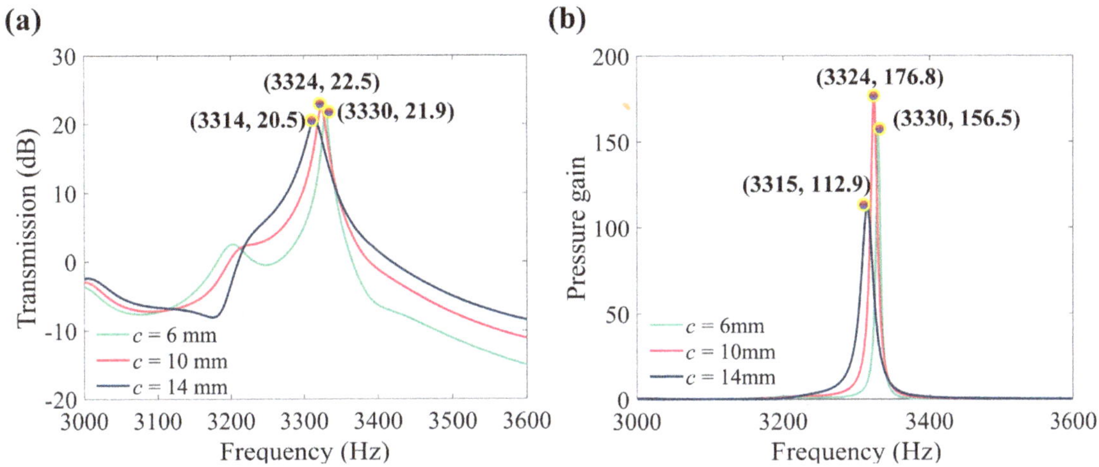

Figure 4. (**a**) Pressure transmission power spectrum and (**b**) Pressure gain spectrum with different straight channel widths c.

To further validate the directional acoustic enhancement capability of the structure, the directional response of the designed line-defect PnC was calculated under varying incident angles α at its resonant frequency of 3324 Hz. Figure 5a presents the absolute sound pressure distributions for three incident angles ($\alpha = -25^\circ$, 0°, and 45°). It is observed that at $\alpha = 0^\circ$, the maximum sound pressure is localized at the center of the line-defect PnC. In contrast, waves incident at other angles (e.g., -25° and 45°) exhibit significant attenuation and become undetectable. The normalized sound pressure gain directional response of the structure was calculated and plotted in Figure 5b. The response curve demonstrates a maximum gain at normal incidence ($\alpha = 0^\circ$) with no observable sidelobe interference. It is worth mentioning that compared to PnCR with equivalent functionality (exceeding 15 dB in transmission power and achieving over 35 times sound pressure gain) [31], the structure designed in this study has a transmission power that is 9 dB higher and a sound pressure gain that is 141 times higher, without the need for complex coupled resonators. This advantage stems from the simplified structural design, which reduces the thermoviscous dissipation of acoustic energy, thereby effectively addressing the high complexity issue commonly faced by high-performance defective PnCs. These theoretical results confirm that the designed line-defect PnC possesses exceptional directional acoustic enhancement properties, establishing it as a high-gain directional acoustic sensing structure.

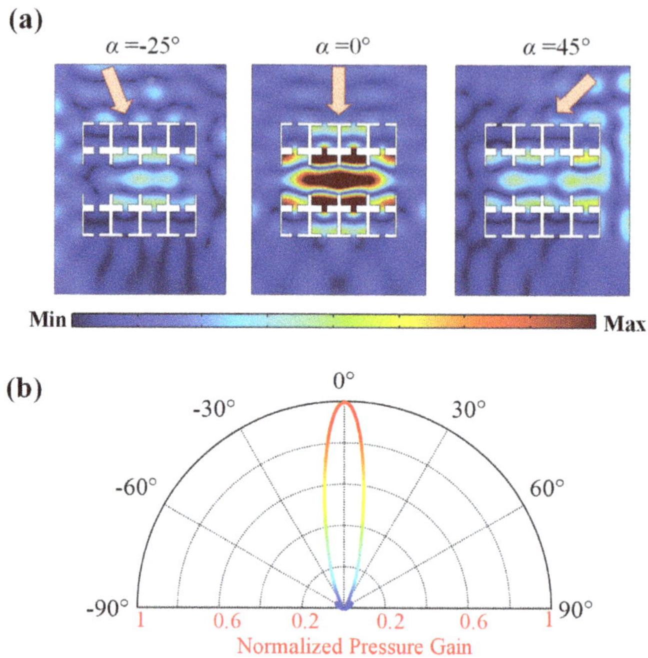

Figure 5. (**a**) Absolute sound pressure distribution of the designed line-defect PnC under plane waves at different incident angles. (**b**) Directional response of normalized pressure gain at resonant frequency 3324 Hz.

3. Analysis of Numerical Simulation Verification Results

3.1. Directional Enhancement Detection of Weak Harmonic Signals

Effectively detecting harmonic signals obscured by spatial noise represents a critical requirement in practical engineering applications. To evaluate the directional enhancement capability of the structure, we constructed a harmonic signal $S_h(t)$ with three distinct incident directions, the specific parameters are shown in Table 1. This was employed to simulate and validate the capability of the structure to detect weak target acoustic signals from the desired direction ($\alpha = 0°$), which is mathematically expressed as follows:

$$S_h(t) = \sum_{n=1}^{N} P_n \sin[2\pi f_n t - k_n(x \sin \alpha_n - y \cos \alpha_n)] \quad (8)$$

Table 1. Detailed parameters of the harmonic signal $S_h(t)$ for numerical simulation detection.

Parameter	Value	Parameter	Value	Value	Parameter
f_n	[3074, 3324, 3674] Hz	α_n	$[45°, 0°, -25°]$	Sampling frequency	8000 Hz
P_n	[1, 10] Pa	N	3	Simulated time	0.3 s

Figure 6a,b show the time-domain and spectrum results of the incident harmonic signal $S_h(t)$ propagating in free space, respectively. Analysis reveals that compared to interference signals from other incident directions, the amplitude of the target sound source at 3324 Hz in the desired direction is extremely weak. However, when detected using the designed line-defect PnC structure, as shown in Figure 6c, the weak harmonic component in the target direction is amplified from an initial 1 Pa to 36 Pa due to the directional enhancement property of this structure. Concurrently, as shown in Figure 6d, interference signals from the other two directions, falling within the structure's bandgap range, experienced significant suppression of their propagated energy. Their amplitudes decreased from 10 Pa to 8 Pa and 3 Pa, respectively. Additionally, to clarify the distinction between this anisotropic property and the bandgap characteristic, this study constructed a signal $S_g(t)$ with the same frequency as $S_h(t)$ but with an incident angle of 0°. This signal was then detected by the line-defect PnC, with the results shown in Figure 6e,f. Comparing Figure 6d,f reveals that for signals with different frequencies but identical propagation directions, the designed line-defect PnC suppresses interference signals through anisotropic directional suppression. Conversely, for signals with both differing frequencies and propagation directions, interference signals are similarly suppressed via anisotropic directional suppression. This distinction not only results in different amplitude characteristics for interference signals but also leads to noticeable differences in their gain performance relative to the target signal. These results demonstrate that even when the target directional sound source is interfered with by signals from other directions, the designed line-defect PnC can effectively detect and recover the weak sound source from the desired direction while simultaneously suppressing irrelevant noise from other directions.

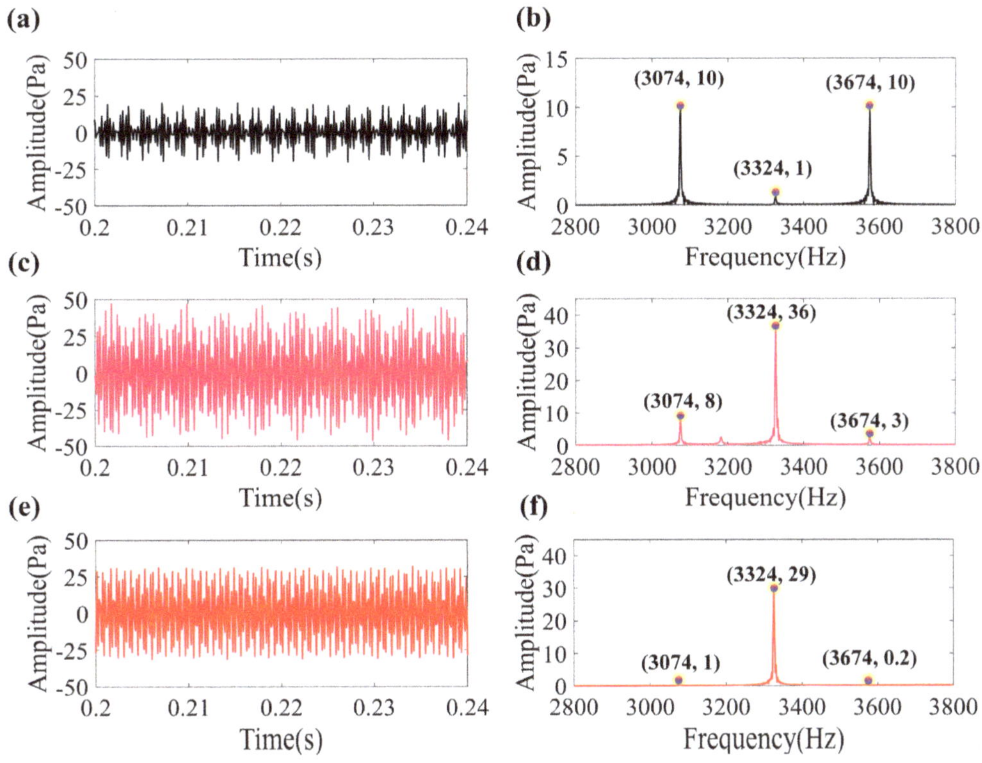

Figure 6. (**a**) Time domain waveform and (**b**) frequency spectrum of the simulation harmonic signal $S_h(t)$ propagating in the free field. (**c**) Time domain waveform and (**d**) frequency spectrum of the simulation harmonic signal $S_h(t)$ propagating in the line-defect PnC. (**e**) Time domain waveform and (**f**) frequency spectrum of the simulation harmonic signal $S_g(t)$ propagating in the line-defect PnC.

3.2. Directional Enhancement Detection of Weak Bearing Fault Signals

Bearings are the most critical fundamental components in rotating machinery, with their operational status directly impacting the reliable functioning of entire mechanical equipment and industrial production systems. When localized bearing failures occur (such as pitting, cracks, or spalling), rolling elements passing through the damaged area collide with it, generating impacts that excite the bearing's natural vibrations and subsequently trigger structural resonance. Consequently, bearing fault signals manifest as amplitude-modulated (AM) signals centered on the carrier frequency with the modulation frequency forming the sidebands. These signals can be approximated as the product of an exponentially decaying function and a sine function [53]:

$$S_b(t) = 2e^{-\varsigma t}\sin(2\pi f_d t) \times [\sin(2\pi f_r t) + 3] \quad (9)$$

where f_r is the rotating frequency, f_n is the natural frequency of the bearing, f_d is the bearing fault frequency, t = mod (t, $0.99/f_n$), and ζ is the damping coefficient. Based on this equation, further simulations were conducted to detect bearing fault signals, verifying the potential of the proposed line-defect PnC in industrial inspection applications. Table 2 lists the parameters of the simulated bearing fault signal $S_b(t)$. Additionally, a strong noise source (maximum amplitude of 6 Pa, uniformly distributed broadband white noise) was positioned at approximately 40° azimuth in the horizontal direction to simulate spatial noise interference on the bearing fault signal, as illustrated in Figure 7a.

Table 2. Detailed parameters of bearing fault signal $S_b(t)$ for numerical simulation detection.

Parameter	Value	Parameter	Value	Parameter	Value
f_r	20 Hz	f_d	12 Hz	Simulated time	0.5 s
f_n	3324 Hz	ζ	100	$f_{sampling}$	8000 Hz

Figure 7b–d displays the time domain diagram, spectrum diagram and envelope spectrum diagram of the simulated bearing signal $S_b(t)$. The bearing fault signal exhibits periodic impacts in the time domain, with its spectrum exhibiting a harmonic cluster centered on the natural frequency f_n. The envelope spectrum obtained via Hilbert transform clearly reveals the characteristic fault frequency fd and its harmonics. Figure 7e–g presents the time domain waveform, spectrum, and envelope spectrum obtained from simulated detection in a free field. Unfortunately, strong noise obscures the bearing fault impact characteristics, rendering the fault sideband signals in the spectrum weak and indistinct. Consequently, the envelope spectrum fails to identify the bearing fault characteristic frequency f_d, resulting in unsuccessful bearing fault detection.

The designed line-defect PnC exhibits outstanding acoustic directional enhancement properties, offering significant advantages for detecting weak bearing fault signals. Figure 8a shows the absolute sound pressure distribution of bearing fault signals under spatial noise interference in the free field. It is evident that no valid information can be captured in the free field under strong noise interference. Figure 8b the corresponding sound pressure distribution of a weak bearing fault signal at the line-defect PnC under spatial noise interference. It is observed that the fault signal energy is concentrated at the center of the line-defect PnC detection point. Surprisingly, the line-defect PnC significantly suppresses noise interference, fully preserving the bearing's fault impact characteristics and faint fault features while markedly amplifying the fault signatures, as shown in Figure 8d,e. Furthermore, comparing Figures 7d and 8e reveals that the envelope spectrum of the online defect PnC detection not only successfully identifies the fault characteristic frequency fd and its harmonics but also exhibits amplitudes far exceeding those of the characteristic frequencies in the bearing fault signal itself. All findings confirm that the proposed line-defect PnC enables directional enhancement of weak sound sources in the target direction amid strong spatial noise interference, while retaining excellent directional gain. This structure thereby demonstrates significant potential for detecting faint acoustic signals.

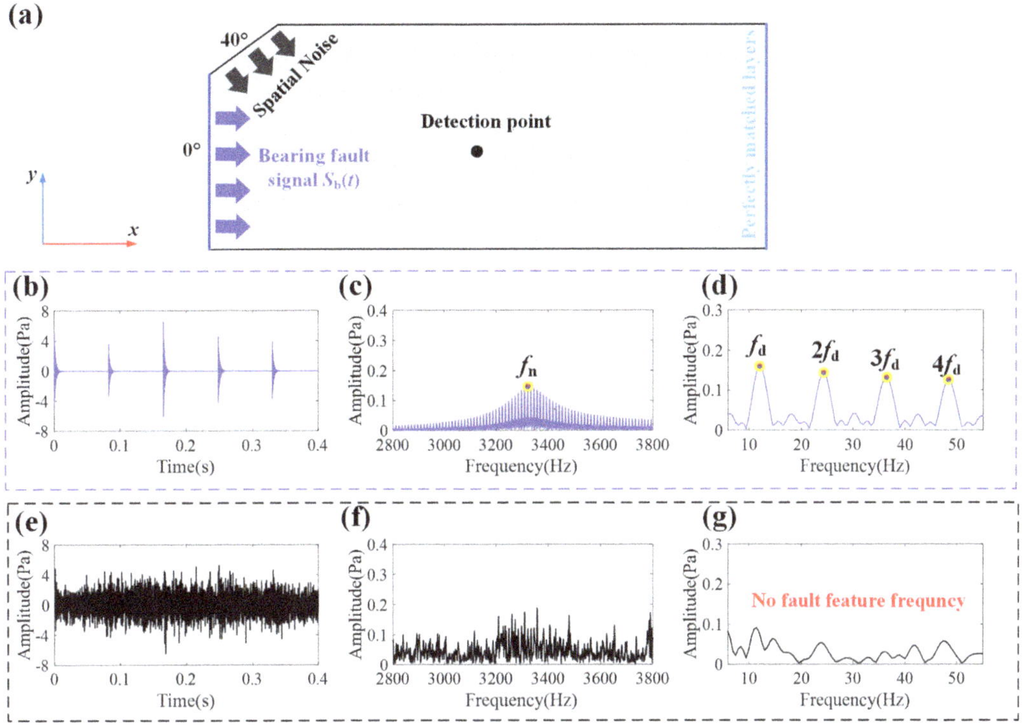

Figure 7. (**a**) Schematic of the propagation of bearing fault signals disturbed by spatial noise in 40° direction. (**b**) Time domain waveform, (**c**) frequency spectrum and (**d**) envelope spectrum diagram of the simulated bearing signal $S_b(t)$. (**e**) Time domain waveform, (**f**) frequency spectrum and (**g**) envelope spectrum diagram of the detected bearing signal $S_b(t)$ in the free field.

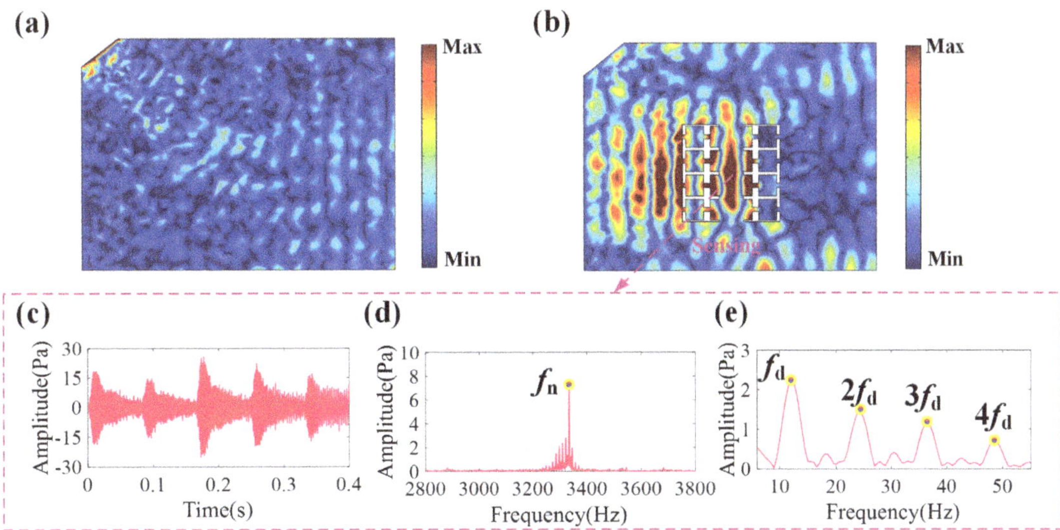

Figure 8. Absolute sound pressure distribution of bearing fault signals under directional noise interference in (**a**) the free field and (**b**) the designed line-defect PnC. (**c**) Time domain waveform, (**d**) frequency spectrum and (**e**) envelope spectrum diagram of the detected bearing signal $S_b(t)$ in the line-defect PnC.

4. Conclusions

To address the technical bottleneck of directional detection of weak acoustic signals in environments with strong spatial noise, this paper proposes a line-defect PnC structure that couples defect-state localization characteristics with structural anisotropy mechanisms. This structure achieves directional enhancement and detection of weak acoustic signals from the target direction. Through theoretical derivation and numerical simulation experiments, the directional enhancement characteristics of the designed structure were systematically evaluated and verified. The main conclusions are as follows:

(1) The proposed PnC exhibits exceptional transmission gain for target signals, theoretically achieving over 24.5 dB acoustic power transmission gain and 176.9 times sound pressure enhancement.

(2) Benefiting from the anisotropic design of the structure, this line-defect PnC achieves outstanding azimuthal resolution without relying on complex coupling structures. It demonstrates high sensitivity and directional sensitivity for target acoustic signals, particularly at points of interest such as 0°. Simultaneously, its straightforward structural design and manufacturing process significantly reduce costs in industrial applications, showcasing remarkable economic advantages.

(3) Compared to conventional acoustic detection techniques, the method based on this line-defect PnC enables directional enhancement and detection of weak acoustic signals without requiring complex array sensor structures or post-processing algorithms. It effectively detects weak harmonic signals and weak bearing fault signals amidst strong spatial noise interference.

In summary, this approach provides novel insights and theoretical guidance for developing high-sensitivity and high-directivity acoustic detection technologies, demonstrating significant application potential in the perception, enhancement, and localization of weak acoustic signals.

Author Contributions: Conceptualization, S.Z. and J.X.; methodology, J.X.; software, S.Z. and J.X.; validation, J.M.; formal analysis, J.X.; investigation, J.M.; resources, J.X.; data curation, J.M.; writing—original draft preparation, S.Z.; writing—review and editing, S.Z. and J.X.; visualization, H.X.; supervision, H.X.; project administration, H.X.; funding acquisition, S.Z. and H.X. All authors have read and agreed to the published version of the manuscript.

Funding: This research was funded by the Key R&D Program of Henan Province, grant number 231111241300, and the Key Scientific Research projects in higher education institutions of Henan Province, grant number 24A460020.

Data Availability Statement: The original contributions presented in this study are included in the article. Further inquiries can be directed to the corresponding author(s).

Acknowledgments: The author sincerely expresses gratitude to all co-authors for their invaluable theoretical contributions and steadfast support throughout this research.

Conflicts of Interest: Huiqiang Xu was employed by China Xuchang Xuji Wind Power Technology Co., Ltd. The remaining authors declare that the research was conducted in the absence of any commercial or financial relationships that could be construed as a potential conflict of interest. The funders had role in the writing of the manuscript and in the decision to publish the results.

References

1. Ivancic, J.; Alves, F. Directional Multi-Resonant Micro-Electromechanical System Acoustic Sensor for Low Frequency Detection. *Sensors* **2024**, *24*, 2908. [CrossRef]
2. Spachos, P.; Gregori, S.; Deen, M.J. Voice Activated IoT Devices for Healthcare: Design Challenges and Emerging Applications. *IEEE Trans. Circuits Syst. II-Express Briefs* **2022**, *69*, 3101–3107. [CrossRef]

3. Zhang, Z.; Wu, M.; Yin, L.; Gong, C.; Wang, J.; Zhou, S.; Yang, J. Robust feedback controller combined with the remote microphone method for broadband active noise control in headrest. *Appl. Acoust.* **2022**, *195*, 108815. [CrossRef]
4. Mohammadgholiha, M.; Zonzini, F.; Moll, J.; De Marchi, L. Directional Multifrequency Guided Waves Communications Using Discrete Frequency-Steerable Acoustic Transducers. *IEEE Trans. Ultrason. Ferroelectr. Freq. Control* **2023**, *70*, 1494–1505. [CrossRef]
5. Diamant, R.; Lampe, L. Low Probability of Detection for Underwater Acoustic Communication: A Review. *IEEE Access* **2018**, *6*, 19099–19112. [CrossRef]
6. Zhang, M.; Li, M.; Xu, W.; Zhang, F.; Yao, D.; Wang, X.; Dong, W. Soft Wireless Passive Chipless Sensors for Biological Applications: A Review. *Biosensors* **2025**, *15*, 6. [CrossRef]
7. He, Y.; Li, M.; Meng, Z.; Chen, S.; Huang, S.; Hu, Y.; Zou, X. An overview of acoustic emission inspection and monitoring technology in the key components of renewable energy systems. *Mech. Syst. Signal Proc.* **2021**, *148*, 107146. [CrossRef]
8. Ding, X.; Li, Y.; Xiao, J.; He, Q.; Yang, X.; Shao, Y. Parametric Doppler correction analysis for wayside acoustic bearing fault diagnosis. *Mech. Syst. Signal Proc.* **2022**, *166*, 108375. [CrossRef]
9. Zhang, S.; He, Q.; Ouyang, K.; Xiong, W. Multi-bearing weak defect detection for wayside acoustic diagnosis based on a time-varying spatial filtering rearrangement. *Mech. Syst. Signal Proc.* **2018**, *100*, 224–241. [CrossRef]
10. Tang, L.; Tian, H.; Huang, H.; Shi, S.; Ji, Q. A survey of mechanical fault diagnosis based on audio signal analysis. *Measurement* **2023**, *220*, 113294. [CrossRef]
11. Li, Z.; Yin, J.; Wei, Y. Study on fault diagnosis of gear fracture based on beamformer. *Appl. Acoust.* **2022**, *199*, 108994. [CrossRef]
12. Huang, S.; Li, J.; Wu, L.; Zhang, W. Research on acoustic fault diagnosis of bearings based on spatial filtering and time-frequency domain filtering. *Measurement* **2023**, *221*, 113533. [CrossRef]
13. Rahaman, A.; Kim, B. An mm-sized biomimetic directional microphone array for sound source localization in three dimensions. *Microsyst. Nanoeng.* **2022**, *8*, 66. [CrossRef]
14. Diaz-Garcia, L.; Reid, A.; Jackson-Camargo, J.C.; Windmill, J.F.C. Toward a Bio-Inspired Acoustic Sensor: *Achroia grisella's* Ear. *IEEE Sens. J.* **2022**, *22*, 17746–17753. [CrossRef]
15. Kaina, N.; Lemoult, F.; Fink, M.; Lerosey, G. Negative refractive index and acoustic superlens from multiple scattering in single negative metamaterials. *Nature* **2015**, *525*, 77–81. [CrossRef]
16. Robillard, J.F.; Bucay, J.; Deymier, P.A.; Shelke, A.; Muralidharan, K.; Merheb, B.; Vasseur, J.O.; Sukhovich, A.; Page, J.H. Resolution limit of a phononic crystal superlens. *Phys. Rev. B* **2011**, *83*, 224301. [CrossRef]
17. Yang, X.; Yin, J.; Yu, G.; Peng, L.; Wang, N. Acoustic superlens using Helmholtz-resonator-based metamaterials. *Appl. Phys. Lett.* **2015**, *107*, 5. [CrossRef]
18. Romero-García, V.; Lamothe, N.; Theocharis, G.; Richoux, O.; García-Raffi, L.M. Stealth Acoustic Materials. *Phys. Rev. Appl.* **2019**, *11*, 054076. [CrossRef]
19. Zhan, J.; Mei, Y.; Li, K.; Zhou, Y.; Chen, J.; Ma, Y. Conformal metamaterial coats for underwater magnetic-acoustic bi-invisibility. *Appl. Phys. Lett.* **2022**, *120*, 094104. [CrossRef]
20. Lian, M.; Duan, L.; Chen, J.; Jia, J.; Su, Y.; Cao, T. Acoustic transmissive cloaking with adjustable capacity to the incident direction. *Microsyst. Nanoeng.* **2022**, *8*, 108. [CrossRef] [PubMed]
21. Ma, F.; Huang, Z.; Liu, C.; Wu, J.H. Acoustic focusing and imaging via phononic crystal and acoustic metamaterials. *J. Appl. Phys.* **2022**, *131*, 11103. [CrossRef]
22. Zhang, F.; Tao, W.; Wang, S.; Hu, Q.; Flowers, G.T.; Gaidai, O. Novel phononic-crystal-arrayed acoustic metalens for long beam focusing in multi-band. *Appl. Phys. Express* **2023**, *16*, 75503. [CrossRef]
23. Park, J.; Lee, G.; Lee, D.; Kim, M.; Rho, J. Double-Focusing Gradient-Index Lens with Elastic Bragg Mirror for Highly Efficient Energy Harvesting. *Nanomaterials* **2022**, *12*, 1019. [CrossRef]
24. Lee, S.; Choi, W.; Park, J.W.; Kim, D.; Nahm, S.; Jeon, W.; Gu, G.X.; Kim, M.; Ryu, S. Machine learning-enabled development of high performance gradient-index phononic crystals for energy focusing and harvesting. *Nano Energy* **2022**, *103*, 107846. [CrossRef]
25. Xiao, J.; Ding, X.; Pan, H.; Zhang, Y.; He, Q.; Shao, Y. Dual-band filtering and enhanced directional via tunable acoustic metamaterial antennas. *Smart Mater. Struct.* **2024**, *33*, 55015. [CrossRef]
26. Zhang, W.; Xin, F. Coiled-up structure with porous material lining for enhanced sound absorption. *Int. J. Mech. Sci.* **2023**, *256*, 108480. [CrossRef]
27. Zhu, X.; Lau, S.; Lu, Z.; Jeon, W. Broadband low-frequency sound absorption by periodic metamaterial resonators embedded in a porous layer. *J. Sound Vib.* **2019**, *461*, 114922. [CrossRef]
28. Bai, Y.; Chen, Y.; Zheng, J. Design of phononic crystal for enhancing low-frequency sound absorption in mufflers. *Sci. Rep.* **2024**, *14*, 28921. [CrossRef]
29. Miniaci, M.; Gliozzi, A.S.; Morvan, B.; Krushynska, A.; Bosia, F.; Scalerandi, M.; Pugno, N.M. Proof of Concept for an Ultrasensitive Technique to Detect and Localize Sources of Elastic Nonlinearity Using Phononic Crystals. *Phys. Rev. Lett.* **2017**, *118*, 214301. [CrossRef]

30. Ma, C.; Gao, S.; Cheng, Y.; Liu, X. Acoustic metamaterial antennas for combined highly directive-sensitive detection. *Appl. Phys. Lett.* **2019**, *115*, 053501. [CrossRef]
31. Xiao, J.; Ding, X.; Wang, Y.; Huang, W.; He, Q.; Shao, Y. Gear fault detection via directional enhancement of phononic crystal resonators. *Int. J. Mech. Sci.* **2024**, *278*, 109453. [CrossRef]
32. Zhang, S.; Hao, G.; Zhao, X.; Liu, Y.; Han, J. Broadband acoustic signal enhancement via gradient metamaterials coupled to crystals. *Front. Phys.* **2023**, *11*, 1240468. [CrossRef]
33. Xiao, J.; Ding, X.; Huang, W.; He, Q.; Shao, Y. Rotating machinery weak fault features enhancement via line-defect phononic crystal sensing. *Mech. Syst. Signal Proc.* **2024**, *220*, 111657. [CrossRef]
34. Yi, C.; Liu, X.; Xiao, C.; Liu, J.; Chen, N. Soft Phononic Crystal with Tunable Bandgap Through Pneumatic Actuation. *Adv. Eng. Mater.* **2024**, *26*, 2401913. [CrossRef]
35. Jo, S. Temperature-Controlled Defective Phononic Crystals with Shape Memory Alloys for Tunable Ultrasonic Sensors. *Crystals* **2025**, *15*, 412. [CrossRef]
36. He, Z.; Zhang, G.; Chen, X.; Cong, Y.; Gu, S.; Hong, J. Elastic wave harvesting in piezoelectric-defect-introduced phononic crystal microplates. *Int. J. Mech. Sci.* **2023**, *239*, 107892. [CrossRef]
37. Wang, K.; Li, X.; Cao, L.; Guo, P.; Fan, G.; Qin, J.; Ma, T. Enhancement of piezoelectric energy harvesting for flexural waves by a metasurface-assisted phononic cavity. *Results Phys.* **2024**, *63*, 107870. [CrossRef]
38. Jiang, T.; He, Q.; Peng, Z. Enhanced directional acoustic sensing with phononic crystal cavity resonance. *Appl. Phys. Lett.* **2018**, *112*, 261902. [CrossRef]
39. Chen, T.; Jiao, J.; Yu, D. Strongly coupled phononic crystals resonator with high energy density for acoustic enhancement and directional sensing. *J. Sound Vib.* **2022**, *529*, 116911. [CrossRef]
40. Zhao, X.; Hao, G.; Shang, Y.; Han, J. Three-Dimensional Gradient Metamaterial Devices Coupled with Phononic Crystals for Acoustic Enhancement Sensing. *Crystals* **2023**, *13*, 1191. [CrossRef]
41. Han, J.; Hao, G.; Yang, W.; Zhao, X. Phononic Crystal Coupled Mie Structure for Acoustic Amplification. *Crystals* **2023**, *13*, 1196. [CrossRef]
42. Lei, Y.; Wu, J.H.; Huang, Z.; Wang, L.; Huang, Y. Broadband directional resonant tunneling emission enhancement via acoustic anisotropic metamaterials. *Appl. Acoust.* **2022**, *200*, 109050. [CrossRef]
43. Qian, J.; Sun, H.; Yuan, S.; Liu, X. Enhanced directional acoustic emission based on anisotropic metamaterials. *Appl. Phys. Lett.* **2019**, *114*, 013506. [CrossRef]
44. Jo, S.; Yoon, H.; Shin, Y.C.; Choi, W.; Park, C.; Kim, M.; Youn, B.D. Designing a phononic crystal with a defect for energy localization and harvesting: Supercell size and defect location. *Int. J. Mech. Sci.* **2020**, *179*, 105670. [CrossRef]
45. Wang, X.; Sun, H.; Chen, T.; Wang, X. Enhanced acoustic localization in the two-dimensional phononic crystals with slit tube defect. *Phys. Lett. A* **2019**, *383*, 125918. [CrossRef]
46. Ma, X.; Li, Z.; Xiang, J.; Wang, C. Optimization of a ring-like phononic crystal structure with bonding layers for band gap. *Mech. Syst. Signal Proc.* **2022**, *173*, 109059. [CrossRef]
47. Li, Y.; Gao, Z.; Cai, K.; Luo, Y. Design of multi-state tunable phononic crystals based on the reconstruction mechanism of guide-rail lattice. *Int. J. Mech. Sci.* **2023**, *254*, 108442. [CrossRef]
48. Jiang, Z.; Zhou, Y.; Zheng, S.; Liu, J.; Xia, B. Waveguides induced by replacing defects in phononic crystal. *Int. J. Mech. Sci.* **2023**, *255*, 108464. [CrossRef]
49. Huang, H.; Huo, S.; Chen, J. Subwavelength elastic topological negative refraction in ternary locally resonant phononic crystals. *Int. J. Mech. Sci.* **2021**, *198*, 106391. [CrossRef]
50. Chen, T.; Wu, B.; Yu, D. Acoustic fault signal extraction via the line-defect phononic crystals. *Front. Mech. Eng.* **2022**, *17*, 10. [CrossRef]
51. Berggren, M.; Bernland, A.; Noreland, D. Acoustic boundary layers as boundary conditions. *J. Comput. Phys.* **2018**, *371*, 633–650. [CrossRef]
52. Thibault, A.; Chabassier, J.; Boutin, H.; Hélie, T. Transmission line coefficients for viscothermal acoustics in conical tubes. *J. Sound Vib.* **2023**, *543*, 117355. [CrossRef]
53. Feng, Z.; Liang, M.; Zhang, Y.; Hou, S. Fault diagnosis for wind turbine planetary gearboxes via demodulation analysis based on ensemble empirical mode decomposition and energy separation. *Renew. Energy* **2012**, *47*, 112–126. [CrossRef]

crystals

MDPI

Article

Tandem Neural Network Based Design of Acoustic Metamaterials for Low-Frequency Vibration Reduction in Automobiles

Jianjiao Deng [1], Jiawei Wu [2], Xi Chen [1], Xinpeng Zhang [1], Shoukui Li [1], Yu Song [1], Jian Wu [1], Jing Xu [1], Shiqi Deng [2] and Yudong Wu [2,3,*]

1 State Key Laboratory of Advanced Vehicle Integration and Control, China FAW Group Co., Ltd., Changchun 130013, China; dengjianjiao@faw.com.cn (J.D.); chenxi3@faw.com.cn (X.C.); zhangxinpeng5@faw.com.cn (X.Z.); lishoukui@faw.com.cn (S.L.); songyu1@faw.com.cn (Y.S.); wujian24@faw.com.cn (J.W.); xujing1@faw.com.cn (J.X.)

2 College of Mechanical Engineering, Southwest Jiaotong University, Chengdu 610031, China; 2024210556@my.swjtu.edu.cn (J.W.); shiqi_deng@yeah.net (S.D.)

3 Key Laboratory of New Energy Vehicles Intelligent Design and Manufacturing in Universities of Sichuan Province, Chengdu 611730, China

* Correspondence: ydwu@swjtu.edu.cn

Abstract: Automotive NVH (Noise, Vibration, and Harshness) performance significantly impacts driving comfort and traffic safety. Vehicles exhibiting superior NVH characteristics are more likely to achieve consumer acceptance and enhance their competitiveness in the marketplace. In the development of automotive NVH performance, traditional vibration reduction methods have proven to be mature and widely implemented. However, due to constraints related to size and weight, these methods typically address only high-frequency vibration control. Consequently, they struggle to effectively mitigate vehicle body and component vibration noise at frequencies below 200 Hz. In recent years, acoustic metamaterials (AMMs) have emerged as a promising solution for suppressing low-frequency vibrations. This development offers a novel approach for low-frequency vibration control. Nevertheless, conventional design methodologies for AMMs predominantly rely on empirical knowledge and necessitate continuous parameter adjustments to achieve desired bandgap characteristics—an endeavor that entails extensive calculations and considerable time investment. With advancements in machine learning technology, more efficient design strategies have become feasible. This paper presents a tandem neural network (TNN) specifically developed for the design of AMMs. The trained neural network is capable of deriving both the bandgap characteristics from the design parameters of AMMs as well as deducing requisite design parameters based on specified bandgap targets. Focusing on addressing low-frequency vibrations in the back frame of automobile seats, this method facilitates the determination of necessary AMMs design parameters. Experimental results demonstrate that this approach can effectively guide AMMs designs with both speed and accuracy, and the designed AMMs achieved an impressive vibration attenuation rate of 63.6%.

Keywords: acoustic metamaterials; vibration control; bandgap characteristics; tandem neural network; inverse design

1. Introduction

The doors, seats, and roofs of automobiles are prone to low-frequency resonance, resulting in significant vibration and noise. These low-frequency vibrations not only affect driving comfort, but can also lead to premature component failure, increasing safety hazards. The bandgap in AMMs usually refers to the fact that within a certain frequency range of periodic materials, the material strongly suppresses the propagation of vibration waves or sound waves, resulting in the inability of waves within this frequency band to propagate within the material. That is, in the bandgap frequency range, vibrations or sound waves are effectively blocked or attenuated. Traditional damping materials are limited by their mass and load-bearing capacity, and it is difficult to achieve the performance of lightweight, high-strength, and low-frequency broadband bandgap at the same time [1,2]. Traditional structural vibration isolators are generally used for medium and high frequency vibration suppression, which has certain limitations.

AMMs [3,4] are a class of structural materials that are artificially designed to have acoustic properties that are not found in natural materials. Therefore, AMMs show great potential in noise reduction [5–9], vibration damping [10,11], energy harvesting [12,13], acoustic Metasurfaces [14–16], and directional wave propagation [17–19]. The use of acoustic metamaterials for low-frequency vibration reduction has been a hot topic of research for various research groups in the field of metamaterials over the years. Ping He [20] proposed a new type of composite acoustic metamaterial to address the low-frequency vibrations below 0–100 Hz generated by in-orbit rotating mechanical equipment. Lixia Li [21] proposed an acoustic black hole radial elastic metamaterial (AREM). The AREM accumulates energy at the tip of the black hole unit, which can trigger the local resonance (LR) effect and couple with the Bragg scattering (BS) effect, thereby opening up a strongly attenuating broadband at extremely low frequencies. Jinchen Zhou [22] utilized the so-called "trampoline effect" to design a low-frequency, broadband negative Poisson's ratio structural acoustic metamaterial (NPRS-SC). The first low-frequency band gap range of NPRS-SC is 66.1 to 281.1 Hz; compared with traditional structures, it has a lower starting frequency and a wider bandgap. Although the application of AMMs has become very widespread, there are still certain thresholds. This is because the relationship between the parameter space of AMMs and the target bandgap is highly nonlinear. Traditional finite element methods (FEM) can usually only perform forward calculations, that is, calculate the target bandgap from the parameters. FEM not only requires the designer to have significant experience, but also must invest a lot of time and energy to constantly try and make mistakes and adjustments in order to finally obtain the suitable target bandgap. However, the problem of inverse design, that is, obtaining the parameters of the AMMs we need from the target bandgap, is difficult at present. Generally speaking, the methods of inverse design include gradient optimization, evolutionary algorithm, and topology optimization, but with the increase of structural complexity, the parameter space also shows geometric multiplication, and the time spent also increases greatly. Therefore, the choice of AMMs design method is crucial.

Machine learning (ML) is a technology that allows computers to automatically learn patterns from data and make predictions or decisions, driving innovation in structural design and continuous technological advancement. In recent years, deep learning (DL), as a branch of machine learning, has been studied in materials science [23], chemistry [24], and physics [25] because of its excellent performance in dealing with complex pattern recognition tasks. With the continuous development of machine learning technology, researchers have been influenced by this technology and have begun to pay attention to the parameter design and performance prediction of materials. The most important advantage of deep learning is its ability to automatically extract high-level features from large amounts of complex data and make effective representations, which can find potential connections between AMMs parameters and performance, and is widely used in AMMs research [26–29]. Specifically, the machine learning methods used by researchers in AMMs include multilayer perceptron (MLP) [30,31], convolutional neural network (CNN) [32,33], tandem neural network (TNN) [34,35], generative adversarial network (GAN) [36,37], variational autoencoder (VAE) [38], and reinforcement learning (RL) [39,40], all of which greatly improve the design efficiency of AMMs.

At present, there are various machine learning methods in AMMs design. The frequently used methods include MLP, CNN, TNN, GAN, VAE, and RL. MLP is suitable for general supervised learning tasks, but it performs poorly in handling high-dimensional data and has non-uniqueness issues in inverse design. Usually, in AMMs research, it also needs to be combined with other algorithms. CNN is highly suitable for processing image data and spatial data and can effectively extract local features in structural design. In AMMs research, it is usually necessary to represent the unit cell structure with binary images. The finer the geometric features of the unit cell structure, the higher the computational cost. Moreover, the manufacturability of the unit cell structure needs to be considered, so the solid regions need to be interconnected. Both VAE and GAN are suitable for generation tasks. VAE has good stability, but it takes a long time to learn and has limited capabilities. Usually, the quality of the generated samples is not as good as that of GAN. GAN can generate high-quality samples, but it requires a large amount of computing resources and data, has complex parameter tuning, and requires a long training time. RL is suitable for solving dynamic decision-making problems, but it has a slow convergence speed, requires a large amount of interaction data, and the modeling design of the environment is complex and prone to getting stuck in local optimal solutions. TNN has high flexibility. In AMMs research, it can both complete forward design and inverse design. In the inverse design of multiple parameters, it can also overcome the non-uniqueness problem and converge quickly. The training time is short, but it requires a large amount of dataset support. This paper focuses on enhancing the inverse design efficiency of AMMs. TNN can converge rapidly in the inverse design of AMMs, with short training time and high accuracy. After training, it can not only quickly obtain the bandgap characteristics from the design parameters of AMMs but also obtain the required design parameters through the bandgap target of AMMs. Even if a large amount of data is needed for support, this is still acceptable.

In this paper, the method of tandem neural network is applied to the study of AMMs. The purpose is to establish the mutual mapping relationship between the design parameters and the bandgap characteristics, then determine the bandgap range from the vibration characteristics measured by the vehicle seat, and obtain the design parameters from the above method and carry out the final experimental verification. The rest of the structure of this paper is as follows: Section 2 describes the finite element calculation method and parametric modeling of AMMs to obtain the sample set. Section 3 introduces the structure and training details of TNN for AMMs design and also verifies the accuracy of TNN. Section 4 uses inverse design network to design AMMs for low-frequency vibration of car seats and install them in car seats for experimental verification. Section 5 summarizes this experimental study.

2. Parametric Modeling and Sample Set of AMMs

2.1. Basic Unit Structure of AMMs

Acoustic metamaterials are generally classified into Bragg Scattering Metamaterials and Local Resonance Metamaterials. Bragg Scattering Metamaterials are based on the Bragg scattering phenomenon of sound waves in periodic structures. Their characteristic is the design of regular periodic structures, which will produce interference effects within a specific frequency range, causing the reflection or transmission of sound waves to be interrupted, thereby forming an acoustic band gap. The working principle of local resonant metamaterials is based on the resonance phenomenon of local structures. These materials usually contain elastic elements or mass elements inside, which can generate resonance at specific frequencies. By ingeniously designing these resonant frequencies, local structures can significantly reduce the propagation of sound waves at specific frequencies. The design of this metamaterial does not require the structure to be periodic but relies on local resonant phenomena within it. The AMMs in this study belong to local resonance metamaterials. The basic unit is composed of a square frame and two local resonance units with the same additional mass. The position of the additional mass is the circular region of the two local resonance units, as shown in Figure 1. When AMMs receive frequency excitation from an external vibration source, local resonant units vibrate, and when the frequency of the elastic wave is close to the resonant frequency of the resonant unit, these local resonant elements absorb and hinder the propagation of sound waves at that frequency, forming a bandgap. In addition, the additional mass at the local resonant unit position can reduce the resonant frequency to achieve vibration control in the low frequency range.

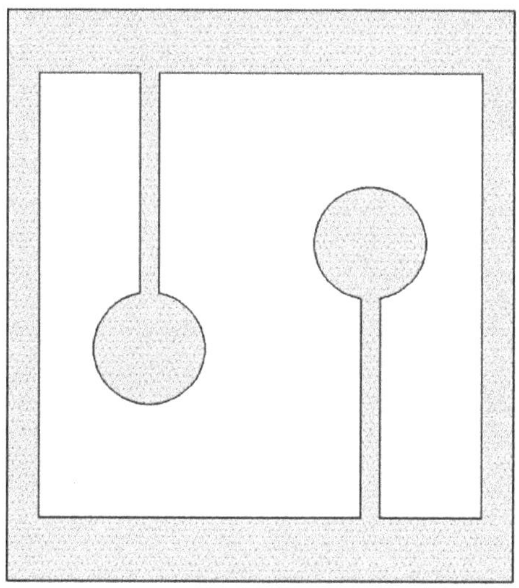

Figure 1. The basic unit of AMMs.

2.2. Bandgap Solution of AMMs

AMMs are arranged in a periodic arrangement, and according to Bloch's theorem, the displacement field of a periodic structure can be expressed as follows:

$$u(r) = U_{\mathrm{k}}(r)e^{i(k\cdot r)} \tag{1}$$

where k is Bloch wave vector, r is position vector, and $u(r)$ is amplitude function. Substituting Equation (1) into the propagation of the governing Equation (2), it can be transformed into the eigenvalue equation of the cell using finite element discretization:

$$\left(K - \omega^2 M\right)u(v) = 0 \tag{2}$$

where K is the structural stiffness matrix and M is the structural mass matrix. Due to periodicity, the wave vector can be limited to the first Brillouin zone, as shown in Figure 2, and the bandgap curve can be accurately obtained by searching for wave vector points ($\Gamma \to X$, $X \to M$, and $M \to \Gamma$) on the Brillouin zone boundary. Considering the computational cost of the training data, only the first four bandgap curves are calculated, and the parameter sweep range of k of the wave vector is 0–3 and the step size is 0.05.

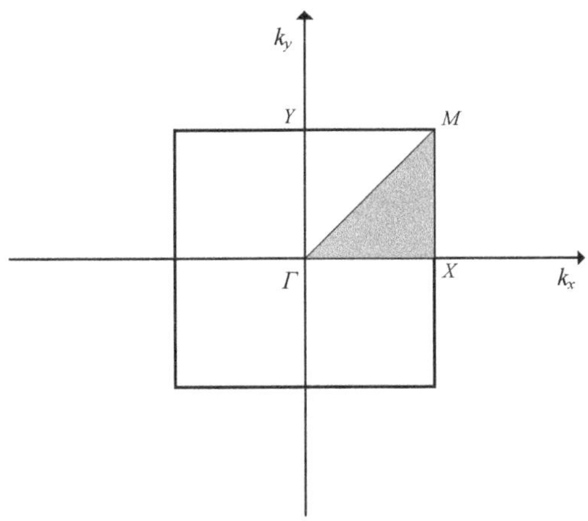

Figure 2. Partition of Brillouin zone.

2.3. Dataset of AMMs with Different Design Parameters

In general, DL requires a sufficient amount of data of good quality to obtain accurate results. Therefore, it is necessary to obtain a dataset with a large number of AMMs samples. In the simulation and experimental tests, the square frame of the AMMs is constrained, that is, the size parameters of the square frame do not affect the resonant frequency of the resonant unit. The parameters that affect the resonant frequency of the basic unit are shown in Figure 3. The four design parameters are the thickness of the cantilever beam (a), the length from the cantilever beam to the center of the circle (b), the radius of the circular area (r), and the additional mass within the area of a single circle (m). The remaining parameters that do not affect the bandgap range are shown in Figure 3, with the unit being millimeters.

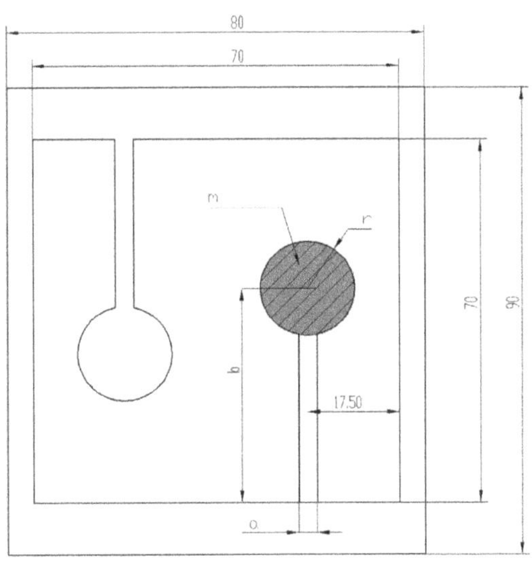

Figure 3. The main parameters of the basic unit affecting the resonant frequency.

In order to avoid interference between the resonant element and the square frame and to obtain high-quality data, the range of variable A is set to 2–10 mm, the range of variable B is set to 15–40 mm, the range of variable r is set to 5–15 mm, and the range is set to 0–100 g of variable m. Using the Latin hypercube (LHS) sampling method, 1024 sets of different parameters were obtained and simulated with COMSOL Multiphysics® 6.2 to extract the upper and lower limits of the bandgap of each model in the band diagram, and the 1024 sets of data were used for deep learning training. This article provides two methods for extracting the upper and lower limits of the band gap of 1024 groups of models. Taking the band gap of AMMs in this paper as an example, the lower limit of the band gap is the frequency when the wave vector of the first dispersion curve is k = 2, and the upper limit of the band gap is the frequency when the wave vector of the second dispersion curve is k = 0.05 or k = 2.95. At this point, the upper and lower limits of the bandgap of 1024 groups of models can be calculated and extracted at one time through COMSOL's parametric scanning function, or the upper and lower limits of the bandgap of each group of models can be extracted through the co-simulation of MATLAB R2024a and COMSOL.

3. AMMs Design Method Based on TNN

3.1. The Framework of TNN for AMMs

This section introduces deep learning frameworks for AMMs design, such as TensorFlow, PyTorch, MXNet, and JAX. PyTorch is known for its "flexibility, ease of use, and active community", which is very suitable for academic research and rapid development of industrial applications, so PyTorch 2.6.0+cpu was used as the software environment in this study.

3.1.1. Forward Pretraining Network

The tandem neural network consists of a pre-trained forward network and an inverse design network, and the structure of the tandem neural network is shown in Figure 4.

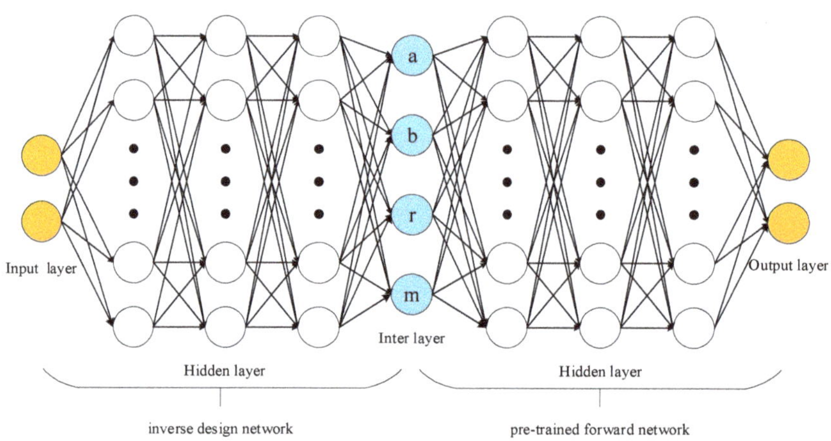

Figure 4. The structure of a tandem neural network.

The pre-trained network is a standard supervised learning network that implements forward mapping of AMMs design parameters to bandgap characteristics. In addition, the pre-trained network after training retains weights and biases that are used to help train the inverse-engineered network. Supervised learning is a machine learning method in which the model is trained on labeled data, meaning that each input sample has a corresponding correct output. During the training process, the model constantly adjusts its own parameters to reduce the difference between the prediction results and the real labels to learn the mapping relationship between input and output. The input layer of the pre-trained network is the design parameters a, b, r, and m, the hidden layer is 3 layers, the number of nodes is 128, 128, and 64, and the output layer is the upper and lower limits of the bandgap.

Before training a neural network, the input data of the training needs to be preprocessed, and the use of preprocessed data can speed up the convergence speed and improve the stability of the training. The methods of data preprocessing are generally standardization or normalization. In this paper, the method of data standardization is adopted. The activation function of the model is the ReLU activation function, and Equation (3) is the mathematical expression of ReLU, which can alleviate the gradient vanishing problem and accelerate the training process. The model uses the mean square error function (MSE) as the loss function, as shown in Equation (4), to quantify the difference between the predicted value and the true value. The adjustment of the learning rate is shown in Equation (5), where α_t is the first learning rate at step t; α_0 is the initial learning rate; γ is the attenuation coefficient ($0 < \gamma < 1$); and t denotes the number of rounds during training. The initial learning rate was 0.001 and was optimized with the Adam optimizer.

$$f(x) = \max(0, x) \tag{3}$$

$$\text{MSE}\left(y, y'\right) = \frac{\sum_{i=1}^{n}\left(y_i - y_i'\right)}{n} \tag{4}$$

$$\alpha_t = \alpha_0 \times \gamma^t \tag{5}$$

3.1.2. Inverse Design Network

The ultimate goal of the inverse design network is to deduce the design parameters from the required frequency bandgap, and its structure is exactly symmetrical to the pre-trained network. In the design process of AMMs, a set of determined design parameters will correspond to a set of determined bandgap ranges, and a set of determined bandgap ranges will have a variety of design parameters corresponding to them, that is, there is a problem of non-unique solution in inverse design. However, the use of pre-trained networks to assist in inverse design of networks can better overcome this problem. The inverse design network usually outputs multiple sets of design parameters during the training process, and these sets of parameters will be evaluated by the pre-trained network, and the pre-trained network will select the one with the lowest loss, that is, the set of design parameters corresponding to the one with the lowest output loss.

The complete training process is shown in Figure 5. The first is the training of the pre-trained network, the specific details of which have been described in detail above and will not be repeated here. It is important to note that the weights and biases in the pre-trained network are no longer updated after the training is completed. The second is the training of the tandem neural network, which consists of an inverse design network and a pre-trained network, and the output of the inverse design network is used as the input of the pre-trained network. It should also be noted that the pre-trained network during the training process of the tandem neural network has been trained before, and the update of weights and biases only occurs in the inverse design network, and the input and output are both bandgap ranges. Finally, after the training of the tandem neural network is completed, the inverse design network is separated from the tandem neural network as an independent neural network model; the weight and bias of the inverse design network are also fixed, and the inverse design network can output design parameters according to the required bandgap range.

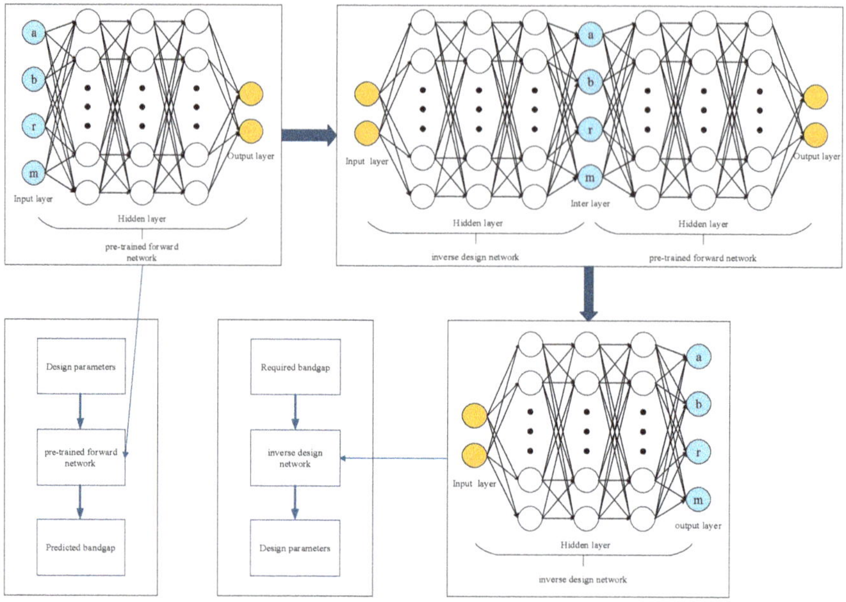

Figure 5. Forward and inverse design flow of AMMs.

3.2. Forward and Inverse Design Results Through the TNN

3.2.1. Forward Design of Bandgap Range Based on Design Parameters

In the pre-trained network, 80% of the 1024 sets of data obtained were divided into training sets, and the remaining 20% were divided into test sets. After multiple iterations of training on the pre-trained network, the loss function is used to measure the gap between the output of the model and the target value so as to evaluate the training effect of the pre-trained network. The training process and all numerical experiments were performed on a Chinese PowerColor AMD Radeon RX 6750 GRE 10 GB GPU and a Chinese AMD Ryzen 5 7500F CPU, but the actual CPU-only version of PyTorch was installed because the GPU did not have CUDA cores. Figure 6a,b represents the training loss function curve and the test loss function curve for the first 1000 iterations of pre-training, respectively. At 200 iterations, it can be found that the loss functions of training and testing converge to a very small value, which indicates that the pre-trained network has been well trained. The total time spent on 1000 iterations of the pre-trained network is 1.22 s, and the average time of one iteration is 0.00122 s.

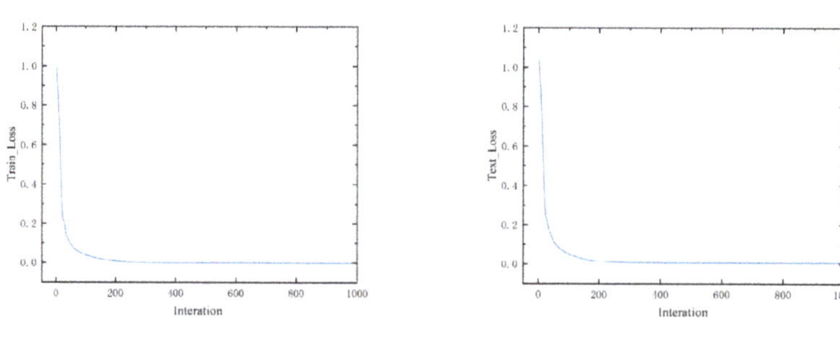

Figure 6. (**a**) Loss function diagram of training set, (**b**) loss function diagram of test set.

The forward design of AMMs can be achieved through a pre-trained network, and the bandgap range can be quickly derived by inputting design parameters. To test the network performance, we randomly input six sets of design parameters and use the pre-trained network and simulation software to obtain the bandgap range. Table 1 shows the comparison of the bandgap results obtained by different methods for these six sets of design parameters, and the error rate is the average error between the predicted upper and lower limits and the upper and lower limits calculated by simulation. Figure 7 shows the upper and lower limits of the bandgap obtained by the network output and simulation calculations. The upper and lower limits of the bandgap obtained by the network are almost the same as those obtained by the simulation calculation, and the error rate is also very low. It can be concluded that the accuracy of the forward prediction of the pre-trained network is very high.

Table 1. Comparison between the network output bandgap and the simulated bandgap under random parameter input.

Sample	a (mm)	b (mm)	r (mm)	m (g)	Network Output (Hz)	Simulation Result (Hz)	Error Rate (%)
1	3.0	36.2	13.5	10.0	60.1–67.0	60.7–67.3	0.72%
2	6.5	25.8	12.0	65.0	65.9–93.0	66.5–93.0	0.45%
3	9.0	29.6	9.5	35.0	85.8–107.9	86.7–109.2	1.11%
4	4.0	42.4	7.0	50.0	30.1–40.2	29.7–40.3	0.80%
5	7.5	31.6	8.0	45.0	64.5–86.2	64.9–85.8	0.48%
6	4.5	18.4	11.0	80.0	84.3–120.7	83.9–120.1	0.49%

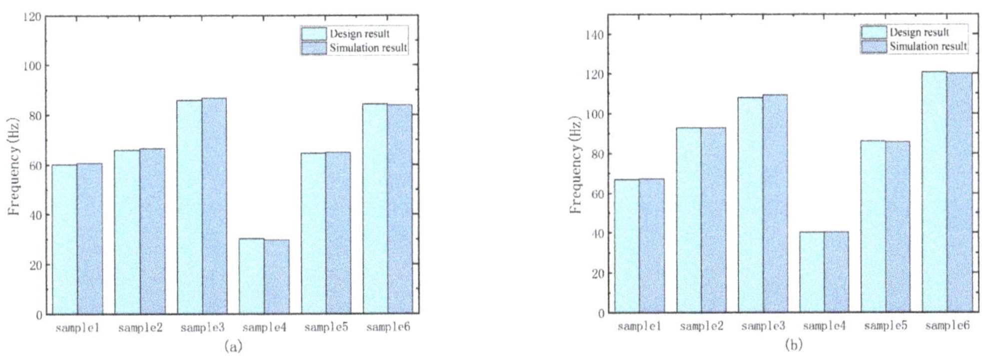

Figure 7. (**a**) Comparison of design lower bound and simulation lower bound; (**b**) Comparison of design upper bound and simulation upper bound.

3.2.2. Inverse Design of Design Parameters Based on Bandgap Range

During the inverse design of the experiment, the loss is defined as the mean square error between the actual bandgap range and the predicted bandgap range in the tandem neural network. During the training process, the best model is obtained by minimizing the bandgap range of the input samples and the bandgap range of the output of the tandem neural network. The training loss function curve is shown in Figure 8, and the tandem neural network has been well trained. The total time spent on training 1000 iterations of a tandem neural network is 1.84 s, and the time for each iteration is 0.00184 s.

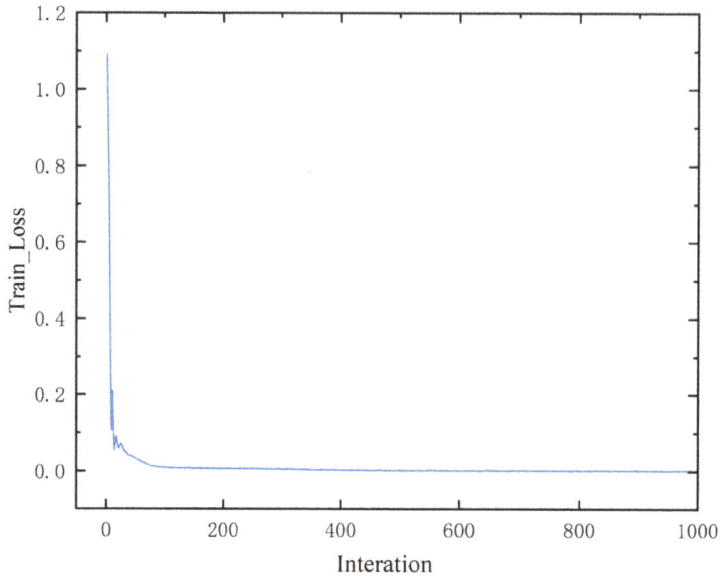

Figure 8. Loss curve of a tandem neural network.

After the training of the tandem neural network is completed, the inverse design network is extracted separately, and the inverse network can output the design parameters of AMMs according to the input bandgap range. The design parameters obtained by the inverse design network were entered into the COMSOL software to obtain the simulation results, and the input bandgap was compared with the simulation results. As shown in Table 2, the difference between the predicted results of the model and the simulation results is small. As shown in Figure 9, the orange portion represents the desired bandgap range, the green portion represents the simulated bandgap range, and the blue portion represents their overlapping portion. The range of the input bandgap is basically consistent with the range calculated by the simulation, which proves the effectiveness of the deep learning method in the design of AMMs.

Table 2. Design parameters obtained by entering the required bandgap.

Input Bandgap Range (Hz)	a (mm)	b (mm)	r (mm)	m (g)	Simulation Result (Hz)	Error Rate (%)
30–40	3.2	42.9	9.2	35.0	30.6–38.9	2.38%
40–55	4.6	38.4	10.2	35.2	42.6–54.1	4.06%
55–70	5.2	34.8	10.4	30.9	55.3–68.6	1.27%
70–90	6.2	30.1	11.0	33.1	71.8–89.6	1.51%
90–110	6.2	27.3	10.4	28.0	89.7–108.9	0.67%
110–130	6.0	25.1	9.1	23.7	109.0–129.2	0.76%

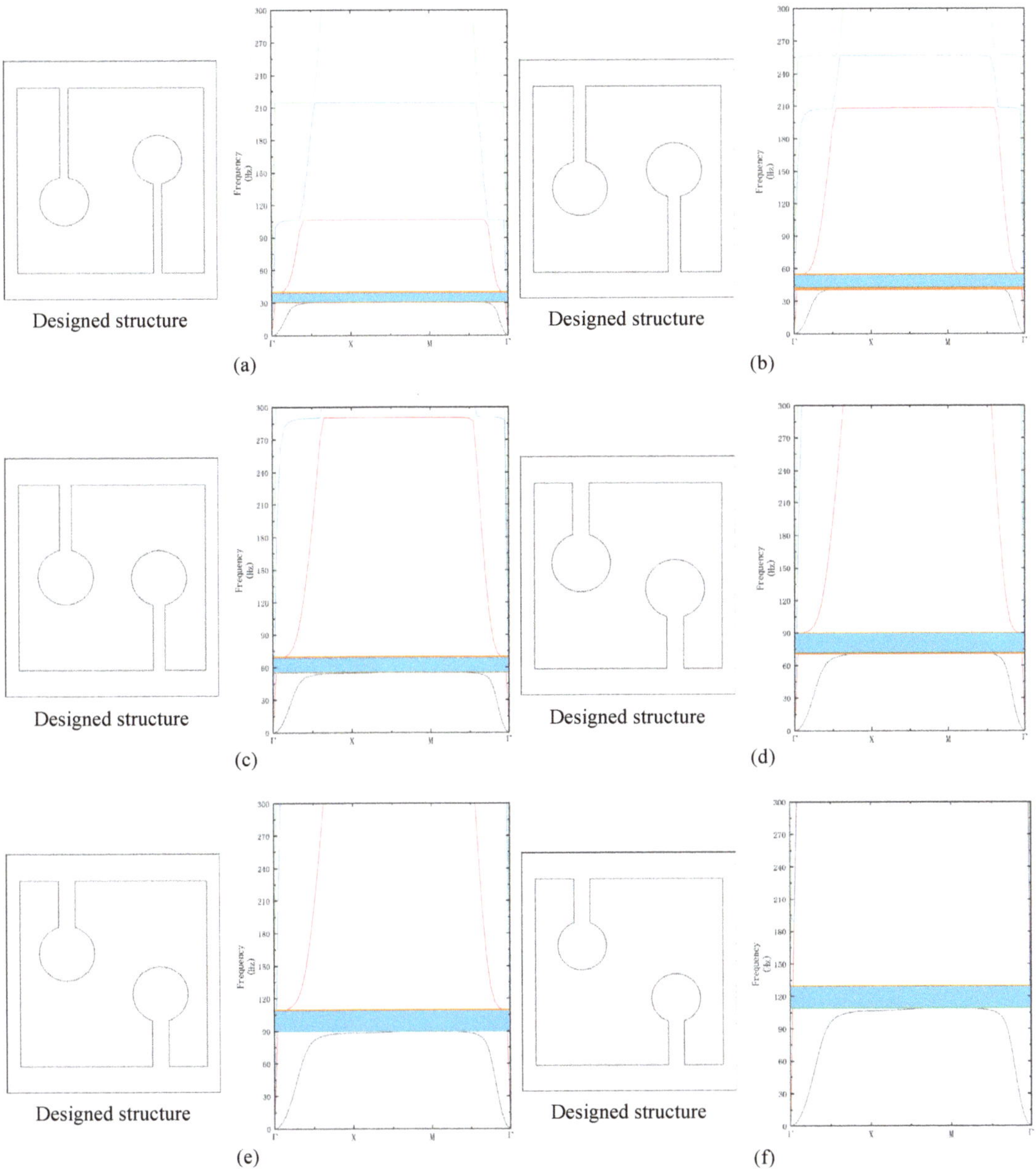

Figure 9. Basic units and corresponding bandgap plots with different structural parameters. (**a**) The input bandgap is 30–40 Hz; (**b**) The input bandgap is 40–55 Hz; (**c**) The input bandgap is 55–70 Hz; (**d**) The input bandgap is 70–90 Hz; (**e**) The input bandgap is 90–110 Hz; (**f**) The input bandgap is 110–130 Hz.

4. Experimental Verification

In the process of driving, due to the influence of the outside world and self-excitation, the doors, seats, roof, and other parts are prone to low-frequency resonance, resulting in obvious vibration and noise. The car seat is an important component that directly contacts the occupants on the entire vehicle and is also a key component that affects the NVH performance of the entire vehicle. The low-frequency vibration and noise of car seats can easily cause driver fatigue, affecting the riding experience and safety.

We selected the problem frequency at the seat back frame of the vehicle to design the AMMs, and we used the inverse design network to design AMMs and experiment on car seats. Figure 10 shows the experimental equipment and measurement scheme, which consists of an LMS system, a hammer, an accelerometer, and a laptop computer. An accelerometer is installed at the seat frame of the vehicle, a hammer is used to tap the position of the rail under the seat, and the frequency response curve in the range of 0–100 Hz is obtained through the LMS system, as shown in Figure 11. We can find that the vibration response of the vehicle seat back frame in the frequency range of 33–40 Hz is large, we set the required bandgap frequency to 32–41 Hz to input the inverse design network, and we get the results shown in Figure 12a, and the bandgap characteristics are shown in Figure 12b. The bandgap range calculated by simulation is 32.5–39.8 Hz.

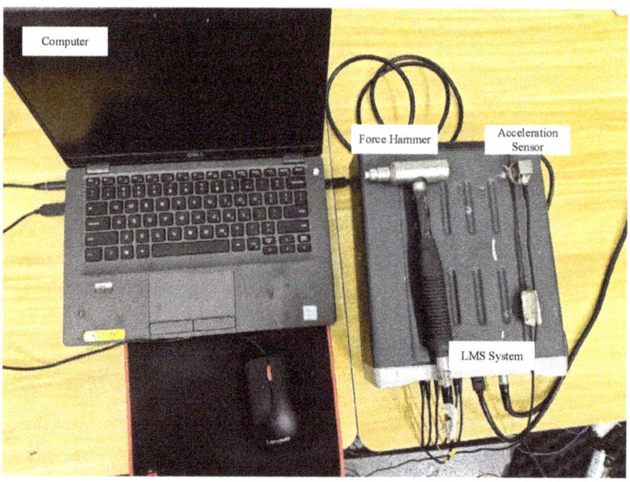

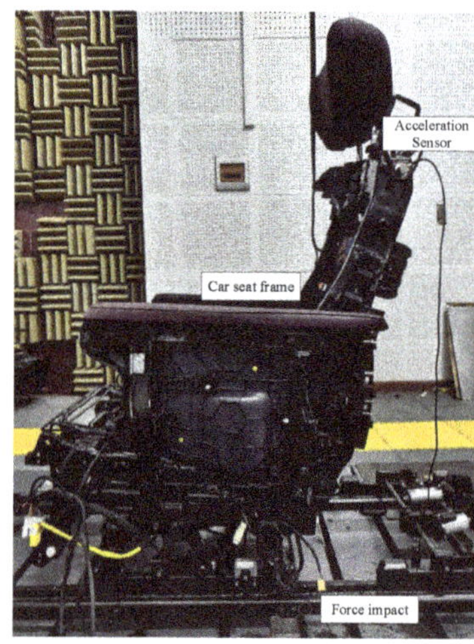

Figure 10. Arrangement of the experimental environment.

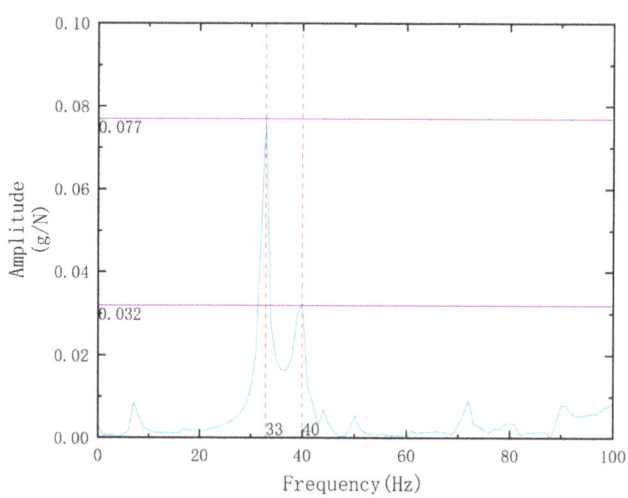

Figure 11. Frequency response curve at the position of the vehicle's seat back frame.

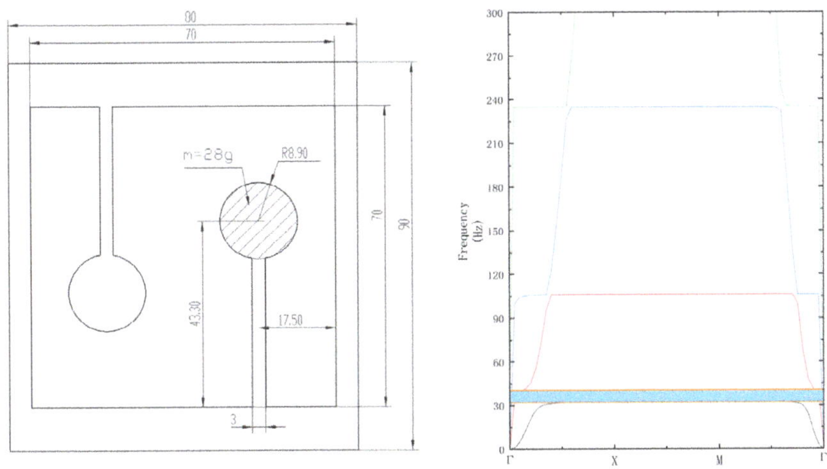

(a) (b)

Figure 12. (**a**) Design parameters for AMMs; (**b**) band diagram of AMMs.

The AMMs in Figure 12a were processed. The processing thickness is 0.9 mm, the material is 304 stainless steel, and the material properties are as follows: the elastic modulus E = 190 GPa, the material density ρ = 7930 kg/m^3 and the Poisson's ratio υ = 0.27. As shown in Figure 13, we arranged three AMMs and mass blocks of the same mass at intervals of 10 cm on a horizontal beam, respectively, and fixed them with cyanoacrylate instant adhesive. The horizontal beam is 50 cm long, 10 cm wide, and 2 cm thick, and is made of steel. Both ends are suspended horizontally by elastic ropes. There is an acceleration sensor on each side of the horizontal beam. The acceleration sensor on the left serves as the response point, and the one on the right serves as the excitation point. A force hammer is used as the excitation source to strike near the right side of the excitation point. After obtaining the sensor signal and the exciter signal through the data acquisition system, the transmission coefficient can be expressed as in Equation (6):

$$T = 20 \times \log_{10}\left|\frac{A_{out}}{A_{in}}\right| \tag{6}$$

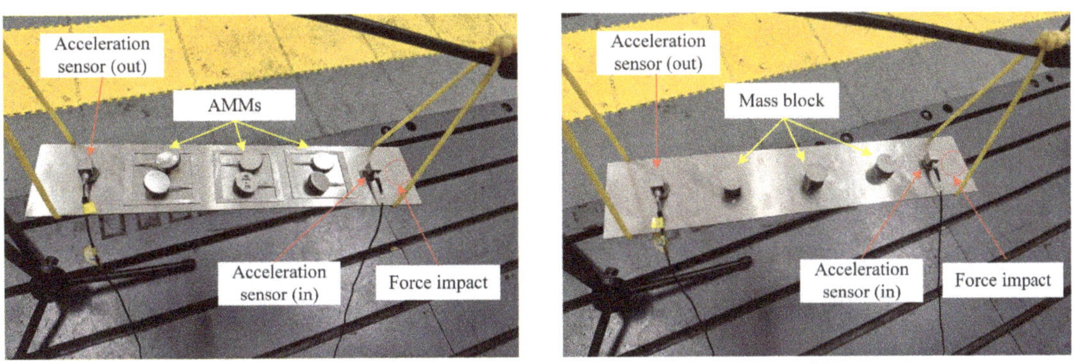

(a) (b)

Figure 13. (**a**) The arrangement of AMMs on the horizontal beam; (**b**) The arrangement of mass blocks on the horizontal beam.

Here, A_{in} and A_{out}, respectively, represent the acceleration amplitudes of the excitation point and the response point.

The transfer coefficients of AMMs and the mass block within the range of 0–100 Hz are shown in Figure 14. The transfer coefficient of the mass block fluctuates around 0 within the range of 30.25–39.75 Hz, while the transfer coefficient of AMMs is less than 0 within the range of 30.25–39.75 Hz. This is because the AMMs in this paper are local resonance metamaterials. When the wave propagates within the range of 30.25–39.75 Hz, the local structure of AMMs resonates with the wave and consumes energy, generating a bandgap in this frequency band. That is, the bandgap range measured by the AMMs experiment is 30.25–39.75 Hz, which has a certain error compared with the bandgap range of 32.5–39.8 Hz calculated by the simulation. This part of the error may be due to the fact that the instant adhesive fixation of cyanoacrylate fails to meet the ideal constraint of the simulation calculation, or it could be the experimental error caused by the certain inclination of the horizontal beam. Although there are some errors between the theoretical and experimental results, the simulation calculation and the experimental results are basically consistent.

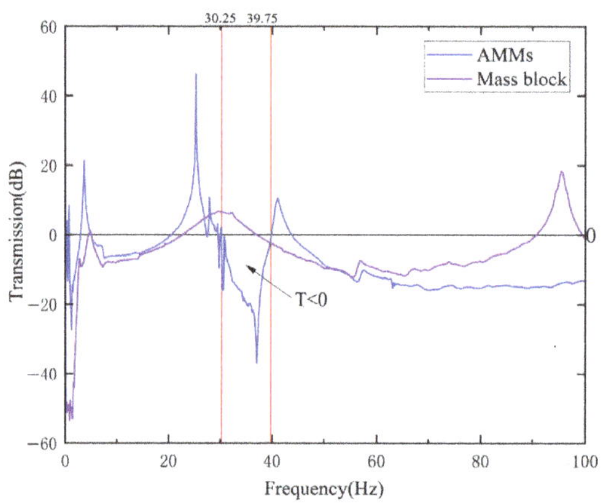

Figure 14. The transfer coefficient between AMMs and mass blocks within the range of 0–100 Hz.

Next, verify the vibration damping effect of AMMs on the car seat back frame. After attaching 28 g circular mass sheets to the AMMs, arrange them on the car seat back frame and fix them with cyanoacrylate instant adhesive. Due to the space limitation of the car seat back frame, multiple AMMs are arranged by bonding three AMMs end to end to the car seat back frame. The specific layout plan is shown in Figure 15.

Once the arrangement is complete, a hammer is used at the striking point to obtain its vibration response in the frequency range of 0–100 Hz. As shown in Figure 16, there is a significant decrease in the vibration response in the frequency range of 32–40 Hz, with a vibration attenuation of about 27.3% at 33 Hz from 0.077 to 0.056 in the case of a single AMMs arrangement, and a decrease of about 63.6% in the case of multiple AMMs in the case of a single AMMs arrangement. The reason for the lower vibration attenuation at 40 Hz may be that the actual upper limit of the bandgap does not fully cover 40 Hz. Overall, however, AMMs do absorb vibrational energy effectively and are in line with the designed bandgap range. The experimental results show that the bandgap characteristics obtained by inverse design are reliable.

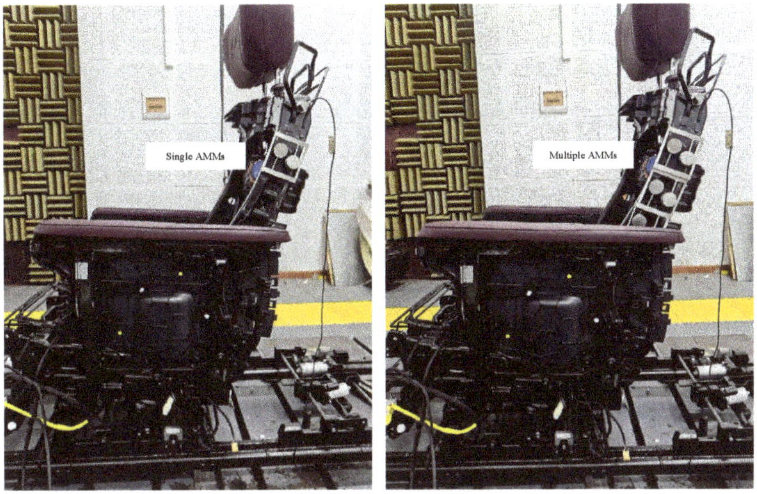

(a) (b)

Figure 15. (**a**) Single AMMS arrangement; (**b**) Multiple AMMS arrangement.

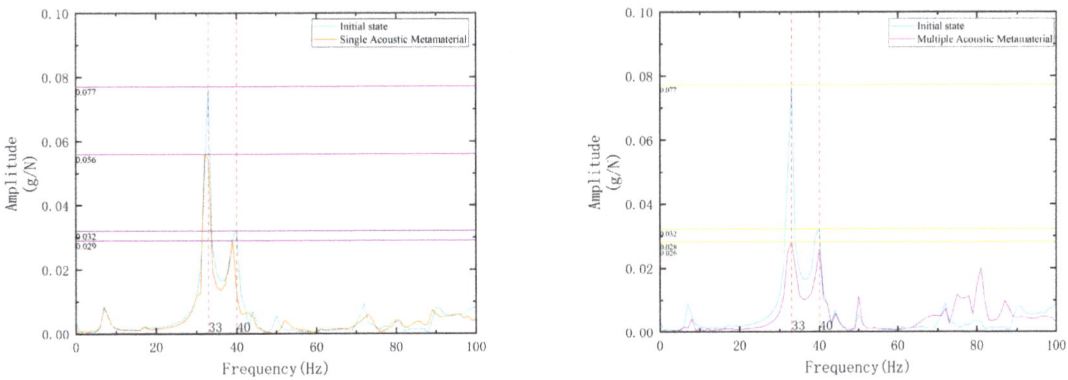

(a) (b)

Figure 16. (**a**) Vibration response under a single AMMs arrangement; (**b**) Vibration response under multiple AMMs arrangement.

5. Conclusions

By introducing the method of TNN to design AMMs, we have designed AMMs for low-frequency vibration of car seats and achieved good results. The conclusions are as follows:

(1) We propose a method to introduce TNN to design AMMs, which can accurately express the mutual mapping relationship between AMMs design parameters and bandgap range. In addition, the forward prediction network in this method can quickly and accurately output the predicted bandgap according to the given design parameters, and the inverse design network can output the design parameters of AMMs according to our desired bandgap range.

(2) We applied this method to the vibration reduction of automobile structures, and designed AMMs for low-frequency vibration of automobile seats and bonded them

to the seat back frame. Finally, the experimental results show that the maximum vibration amplitude is attenuated by 27.3% when single AMMs are pasted, and 63.6% when multiple AMMs are pasted.

The neural network model in this paper can be accurately run within the range of the current four parameters. However, due to the narrow bandgap of AMMs in the low frequency range, if a wide bandgap requirement is given in the process of inverse design, the design parameters of the output may be inaccurate. Despite some problems, this method still has great potential for the design of AMMs, which can be used to design AMMs with multiple bandgaps. In conclusion, the use of tandem neural networks as a deep learning method can effectively improve efficiency and speed up design. In the future, we will try to use this method to assist in the design of multi-mode AMMs, or consider using other intelligent algorithms to improve the accuracy and efficiency of AMMs design.

Author Contributions: Conceptualization, Y.W., J.W. (Jiawei Wu) and S.D.; methodology, J.D., Y.W. and J.W. (Jiawei Wu); software, X.Z. and J.W. (Jian Wu); validation, J.D., X.C. and Y.S.; formal analysis, S.L. and J.X.; investigation, J.D. and X.C.; resources, Y.W. and J.W.; data curation, J.D., Y.S. and J.W. (Jian Wu); writing—original draft preparation, Y.W. and J.W. (Jiawei Wu); writing—review and editing, S.D. and S.L.; visualization, X.Z.; supervision, Y.W.; project administration, S.D.; funding acquisition, Y.W. All authors have read and agreed to the published version of the manuscript.

Funding: This study was funded by the 2023 Open Foundation of the State Key Laboratory of Advanced Vehicle Integration and Control [grant number RBJ33], the Talent Program (Ph.D. Fund) of Chengdu Technological University [grant number 2024RC025] and the Science and Technology Program of Yibin [grant number 2024JC017].

Data Availability Statement: The detailed data supporting the results of this study are available from the corresponding authors upon request.

Conflicts of Interest: Author J.D., X.C., X.Z., S.L., Y.S., J.W. and J.X. were employed by the company State Key Laboratory of Advanced Vehicle Integration and Control, China FAW Group Co., Ltd., Changchun 130013, China. The remaining authors declare that the research was conducted in the absence of any commercial or financial relationships that could be construed as a potential conflict of interest.

References

1. Li, Y.; Zhang, H. Band gap mechanism and vibration attenuation characteristics of the quasi-one-dimensional tetra-chiral metamaterial. *Eur. J. Mech.-A/Solids* **2022**, *92*, 104478. [CrossRef]
2. Wen, Z.; Jin, Y.; Gao, P.; Zhuang, X.; Rabczuk, T.; Djafari-Rouhani, B. Topological cavities in phononic plates for robust energy harvesting. *Mech. Syst. Signal Process.* **2022**, *162*, 108047. [CrossRef]
3. Liu, Z.; Zhang, X.; Mao, Y.; Zhu, Y.Y.; Yang, Z.; Chan, C.T.; Sheng, P. Locally Resonant Sonic Materials. *Science* **2000**, *289*, 1734–1736. [CrossRef]
4. Lu, M.-H.; Feng, L.; Chen, Y.-F. Phononic crystals and acoustic metamaterials. *Mater. Today* **2009**, *12*, 34–42. [CrossRef]
5. Arjunan, A.; Baroutaji, A.; Robinson, J.; Vance, A.; Arafat, A. Acoustic metamaterials for sound absorption and insulation in buildings. *Build. Environ.* **2024**, *251*, 111250. [CrossRef]
6. Wen, G.; Zhang, S.; Wang, H.; Wang, Z.-P.; He, J.; Chen, Z.; Liu, J.; Xie, Y.M. Origami-based acoustic metamaterial for tunable and broadband sound attenuation. *Int. J. Mech. Sci.* **2023**, *239*, 107872. [CrossRef]
7. Zhang, Z.; Wang, X.; Liu, Z.Y.; Fan, Q.; Lin, T.R. A study of low frequency sound insulation mechanism of a perforated plate-type acoustic metamaterial. *J. Sound Vib.* **2023**, *558*, 117775. [CrossRef]
8. Deng, S.; Hu, H.; Qi, Z.; Wu, Y.; Ding, W. An Electrically Tunable Spiral Acoustic Metastructure for Reducing In-Vehicle Noise with Varying Excitation Frequencies. *J. Vib. Eng. Technol.* **2025**, *13*, 322. [CrossRef]
9. Wu, Y.; Yan, W.; Wen, G.; He, Y.; Deng, S.; Ding, W. Design and Application of a Lightweight Plate-Type Acoustic Metamaterial for Vehicle Interior Low-Frequency Noise Reduction. *Crystals* **2024**, *14*, 957. [CrossRef]
10. Lu, K.; Zhou, G.; Gao, N.; Li, L.; Lei, H.; Yu, M. Flexural vibration bandgaps of the multiple local resonance elastic metamaterial plates with irregular resonators. *Appl. Acoust.* **2020**, *159*, 107115. [CrossRef]

11. Yoon, J.Y.; Lim, S.; Yoo, J.; Park, N.-C. Vibration reduction of cables with pendulum-type elastic metamaterials. *Int. J. Mech. Sci.* **2022**, *220*, 107169. [CrossRef]
12. Lu, Z.-Q.; Zhao, L.; Ding, H.; Chen, L.-Q. A dual-functional metamaterial for integrated vibration isolation and energy harvesting. *J. Sound Vib.* **2021**, *509*, 116251. [CrossRef]
13. Ma, T.-X.; Fan, Q.-S.; Li, Z.-Y.; Zhang, C.; Wang, Y.-S. Flexural wave energy harvesting by multi-mode elastic metamaterial cavities. *Extreme Mech. Lett.* **2020**, *41*, 101073. [CrossRef]
14. Assouar, B.; Liang, B.; Wu, Y.; Li, Y.; Cheng, J.-C.; Jing, Y. Acoustic metasurfaces. *Nat. Rev. Mater.* **2018**, *3*, 460–472. [CrossRef]
15. Tian, Z.; Shen, C.; Li, J.; Reit, E.; Gu, Y.; Fu, H.; Cummer, S.A.; Huang, T.J. Programmable Acoustic Metasurfaces. *Adv. Funct. Mater.* **2019**, *29*, 1808489. [CrossRef] [PubMed]
16. Lv, P.; Yang, J.; Huo, C.; Pagliaroli, T. Predicting the ultrasonically suppressive behavior of acoustic metasurfaces. *J. Sound Vib.* **2022**, *535*, 117086. [CrossRef]
17. Beli, D.; Arruda, J.; Ruzzene, M. Wave propagation in elastic metamaterial beams and plates with interconnected resonators. *Int. J. Solids Struct.* **2018**, *139–140*, 105–120. [CrossRef]
18. Zhang, K.; Zhao, P.; Zhao, C.; Hong, F.; Deng, Z. Study on the mechanism of band gap and directional wave propagation of the auxetic chiral lattices. *Compos. Struct.* **2020**, *238*, 111952. [CrossRef]
19. Wang, X.; Li, J.; Wang, Y.; Liu, Y. Active control of wave propagation direction of elastic metamaterial. *Eur. J. Mech.-A/Solids* **2025**, *113*, 105685. [CrossRef]
20. He, P.; Zhang, Y.; Lv, Q.; Wang, Y.; Liu, Z. Research on the Vibration Damping Capabilities of Low-Frequency Bandgap Star-Quadrilateral-Coupled Acoustic Metamaterials. *Arab. J. Sci. Eng.* **2025**, 1–17. [CrossRef]
21. Li, L.; Hu, H.; Wu, X. Ultra-Low-Frequency Acoustic Black Hole Radial Elastic Metamaterials. *Appl. Sci.* **2023**, *13*, 11542. [CrossRef]
22. Zhou, J. Tunable low-frequency wideband acoustic metamaterials with negative Poisson's ratio and pre-compression. *Int. J. Mech. Mater. Des.* **2024**, *20*, 959–972. [CrossRef]
23. Sanchez-Lengeling, B.; Aspuru-Guzik, A. Inverse molecular design using machine learning: Generative models for matter engineering. *Science* **2018**, *361*, 360–365. [CrossRef] [PubMed]
24. Goh, G.B.; Hodas, N.O.; Vishnu, A. Deep learning for computational chemistry. *J. Comput. Chem.* **2017**, *38*, 1291–1307. [CrossRef] [PubMed]
25. Baldi, P.; Sadowski, P.; Whiteson, D. Searching for exotic particles in high-energy physics with deep learning. *Nat. Commun.* **2014**, *5*, 4308. [CrossRef]
26. Muhammad; Kennedy, J.; Lim, C. Machine learning and deep learning in phononic crystals and metamaterials—A review. *Mater. Today Commun.* **2022**, *33*, 104606. [CrossRef]
27. Donda, K.; Brahmkhatri, P.; Zhu, Y.; Dey, B.; Slesarenko, V. Machine learning for inverse design of acoustic and elastic metamaterials. *Curr. Opin. Solid State Mater. Sci.* **2025**, *35*, 101218. [CrossRef]
28. Guo, H.; Chen, W.; Wang, Y.; Ma, F.; Sun, P.; Yuan, T.; Xie, X. Parametric modeling and deep learning-based forward and inverse design for acoustic metamaterial plates. *Mech. Adv. Mater. Struct.* **2024**, *31*, 12986–12996. [CrossRef]
29. Huang, J.; Chen, J.; Mai, H.; Wan, H.; Chen, R.; He, T. Performance prediction and inverse design of cylindrical plate-type acoustic metamaterials based on deep learning. *Appl. Acoust.* **2025**, *234*, 110633. [CrossRef]
30. Jin, Y.; Zeng, S.; Wen, Z.; He, L.; Li, Y.; Li, Y. Deep-subwavelength lightweight metastructures for low-frequency vibration isolation. *Mater. Des.* **2022**, *215*, 110499. [CrossRef]
31. He, L.; Wen, Z.; Jin, Y.; Torrent, D.; Zhuang, X.; Rabczuk, T. Inverse design of topological metaplates for flexural waves with machine learning. *Mater. Des.* **2021**, *199*, 109390. [CrossRef]
32. Jiang, W.; Zhu, Y.; Yin, G.; Lu, H.; Xie, L.; Yin, M. Dispersion relation prediction and structure inverse design of elastic metamaterials via deep learning. *Mater. Today Phys.* **2022**, *22*, 100616. [CrossRef]
33. Donda, K.; Zhu, Y.; Merkel, A.; Wan, S.; Assouar, B. Deep learning approach for designing acoustic absorbing metasurfaces with high degrees of freedom. *Extreme Mech. Lett.* **2022**, *56*, 101879. [CrossRef]
34. Liu, C.-X.; Yu, G.-L.; Zhao, G.-Y. Neural networks for inverse design of phononic crystals. *AIP Adv.* **2019**, *9*, 085223. [CrossRef]
35. Li, J.; Miao, Z.; Li, S.; Ma, Q. Inverse Design of Micro Phononic Beams Incorporating Size Effects via Tandem Neural Network. *Materials* **2023**, *16*, 1518. [CrossRef] [PubMed]
36. Xiao, L.; Cao, Z.; Lu, H.; Cai, Y. Controllable and scalable gradient-driven optimization design for two-dimensional metamaterials based on deep learning. *Compos. Struct.* **2024**, *337*, 118072. [CrossRef]
37. Zhou, H.; Chen, N.; Xia, B.; Man, X.; Liu, J. A data-driven inverse design framework for tunable phononic crystals. *Eng. Struct.* **2025**, *327*, 119599. [CrossRef]
38. Mahesh, K.; Ranjith, S.K.; Mini, R.S. A deep autoencoder based approach for the inverse design of an acoustic-absorber. *Eng. Comput.* **2024**, *40*, 279–300. [CrossRef]

39. Luo, C.; Ning, S.; Liu, Z.; Zhuang, Z. Interactive inverse design of layered phononic crystals based on reinforcement learning. *Extrem. Mech. Lett.* **2020**, *36*, 100651. [CrossRef]
40. He, L.; Guo, H.; Jin, Y.; Zhuang, X.; Rabczuk, T.; Li, Y. Machine-learning-driven on-demand design of phononic beams. *Sci. China Phys. Mech. Astron.* **2022**, *65*, 214612. [CrossRef]

crystals

MDPI

Article

Design of Multifunctional Polarization Waveplates Based on Thermal Phase-Change Metasurfaces

Bo Cheng [1,2], Yuxiao Zou [3], Zihui Ge [4], Longfeng Lv [4], Taohua Liang [1,2,*], Kunpeng Zhai [5,*] and Guofeng Song [4,*]

1 Postdoctoral Innovation Practice Base, Chengdu Polytechnic, 83 Tianyi Street, Chengdu 610041, China
2 Sichuan Provincial Engineering Research Center of Thermoelectric Materials and Devices, Chengdu 610041, China
3 Kunming Institute of Physics, Kunming 650223, China
4 Institute of Semiconductors, Chinese Academy of Sciences, Beijing 100083, China
5 Institute of Intelligent Photonics, Nankai University, Tianjin 300071, China
* Correspondence: lthua222@163.com (T.L.); kpzhai@nankai.edu.cn (K.Z.); sgf@semi.ac.cn (G.S.)

Abstract: The switching function of traditional waveplates necessitates mechanical replacement or the superimposition of multiple waveplates, which gives rise to a complex system and a large volume. We have devised a multifunctional micro-waveplate based on the COMSOL simulation platform (v5.6), which concurrently integrates the compact nature of metasurfaces and the dynamic regulatory features of phase-change materials. When the phase-change material is in the crystalline phase, the metasurface possesses the functionality of a half-waveplate (HWP) and is capable of performing chirality inversion of circularly polarized light within the wavelength range of 1.45 μm to 1.52 μm and 1.56 μm to 1.61 μm. When the phase-change material is in the amorphous phase, the metasurface serves as a quarter-waveplate (QWP) and can achieve the conversion between linear and circular polarization through a 90° phase delay. The phase-change metasurface breaks through the constraint of fixed functions of traditional optical waveplates, facilitating the development of optical systems towards miniaturization, intelligence, and low power consumption and providing a crucial technical route for the next generation of photonic integration and dynamic optical applications.

Keywords: metasurface; waveplates; phase-change materials

1. Introduction

Traditional waveplates (such as quartz, mica, or polymer waveplates) are fundamental polarization control components in optical systems [1]. Through the thickness design of birefringent materials [2,3] (such as quartz), the functional regions of waveplates can cover the ultraviolet to infrared bands. Taking reflective waveplates as an example, in the case of 45-degree linearly polarized light incidence, the amplitudes of the projected components of the reflected light in the x-axis and y-axis directions will be equal, and the phase difference will typically be 90 degrees (corresponding to the QWP [4]) and 180 degrees (corresponding to the HWP [5]). Traditional waveplates possess advantages such as stable optical performance, high phase retardation precision, and low influence from the environment (temperature, humidity) [6]. Nevertheless, they also encounter drawbacks such as fixed functions, volume and weight limitations, and low design freedom. In addition, the photoelastic modulator (PEM) is also a powerful tool for achieving phase delay, and it can function over a wide range of wavelengths. However, it cannot be ignored that it still faces the predicament of being expensive at present.

Metasurfaces [7–11] are two-dimensional planar optical devices composed of subwavelength structures (such as nanoantennas and resonators), capable of flexibly manipulating the amplitude, phase, polarization, and other properties of light waves. Through ultrathin structures, multifunctional integration, and low-cost manufacturing, metasurface technology has significantly improved the performance and application range of optical waveplates. Over the past decade or so, a large number of outstanding metasurface structure designs have emerged, with a focus on optimizing the efficiency [12–16] and bandwidth [17–21] of compact optical waveplates. Phase-change metasurfaces [22,23] are intelligent optical devices that combine phase-change materials (PCMs) and metasurfaces and can dynamically regulate the multi-dimensional characteristics of light waves through external stimuli (such as heat [24], light [25], and electricity [26]). By designing waveplates using phase-change metasurfaces, one can not only possess the ultrathin structural characteristics of metasurfaces but also overcome the defect of the single functionality of traditional waveplates.

In this paper, we have designed an optical metasurface based on the dynamic regulation property of the phase-change material Sb_2Se_3 [27,28]. The activation and quenching of the phase-change material are proposed to be accomplished by employing thermoelectric cooler (TEC) technology [29]. When Sb_2Se_3 is in the amorphous state, the metasurface possesses the functionality of a quarter-waveplate, which can transform linearly polarized light into circularly polarized light. When Sb_2Se_3 is in the crystalline state, the metasurface presents the characteristic of a half-waveplate, allowing for the rotation of the polarization direction of linearly polarized light or the chirality inversion of circularly polarized light. Furthermore, we have analyzed the attenuation effect of potential process errors on the performance of the metasurface.

2. Materials and Methods

Sb_2Se_3 is a typical phase-change material, presenting significant disparities in optical properties (such as refractive index, extinction coefficient, band gap, etc.) between the amorphous (a-Sb_2Se_3) and crystalline (c-Sb_2Se_3) states. Figure 1a depicts the functional diagram of the phase-change metasurface based on Sb_2Se_3. The metasurface is composed of a TEC, a metal reflector, and a periodic array of Sb_2Se_3 cubic cells from bottom to top. Figure 1b shows the three-dimensional structure of the TEC, which mainly comprises N-type and P-type semiconductor thermocouple pairs (BiTe), metal current collectors, a ceramic substrate, and electrodes. Based on the Peltier effect [30,31], energy transfer takes place at the junction of N-type and P-type semiconductors. When a direct current passes through the TEC in the forward direction, it absorbs heat (cooling effect), while in the reverse direction, the TEC releases heat (heating effect), thereby achieving active heat transfer from the cold end to the hot end, resulting in cooling or heating outcomes. When electrons flow from the P-type to the N-type semiconductor, the top surface of the TEC is the cold side, and Sb_2Se_3 is in the amorphous state. The metasurface demonstrates the functionality of a quarter-waveplate, which can convert linearly polarized light into circularly polarized light. When the current reverses, the top surface of the TEC becomes the hot side, and the metasurface with crystalline Sb_2Se_3 behaves as a half-waveplate. Figure 1c,d describe the material attributes and structural dimensions of the metasurface. The optical refractive index parameters of Sb_2Se_3 in the two distinct states are derived from Reference [32]. The refractive index data of the gold are sourced from Reference [33]. The optical model for calculating the efficiency and phase of the metasurface's reflection spectrum is established on the COMSOL platform. Periodic boundary conditions are applied to the four lateral interfaces to converge the computational domain, and perfect matching layers are added to the top and bottom interfaces. The phase and amplitude information of the incident and reflected light are obtained through port modes. The thickness $h2$ of the gold reflector is significantly greater than the skin depth of the gold (~20 nm), ensuring the absence of photons in the transmission domain. Furthermore, only a gold film with a thickness of 600 nm needs to be grown on the ceramic substrate of the TEC, followed by physical deposition of an Sb_2Se_3 film. After the electron beam lithography process for patterning the electron glue and etching, the metasurface can be fabricated.

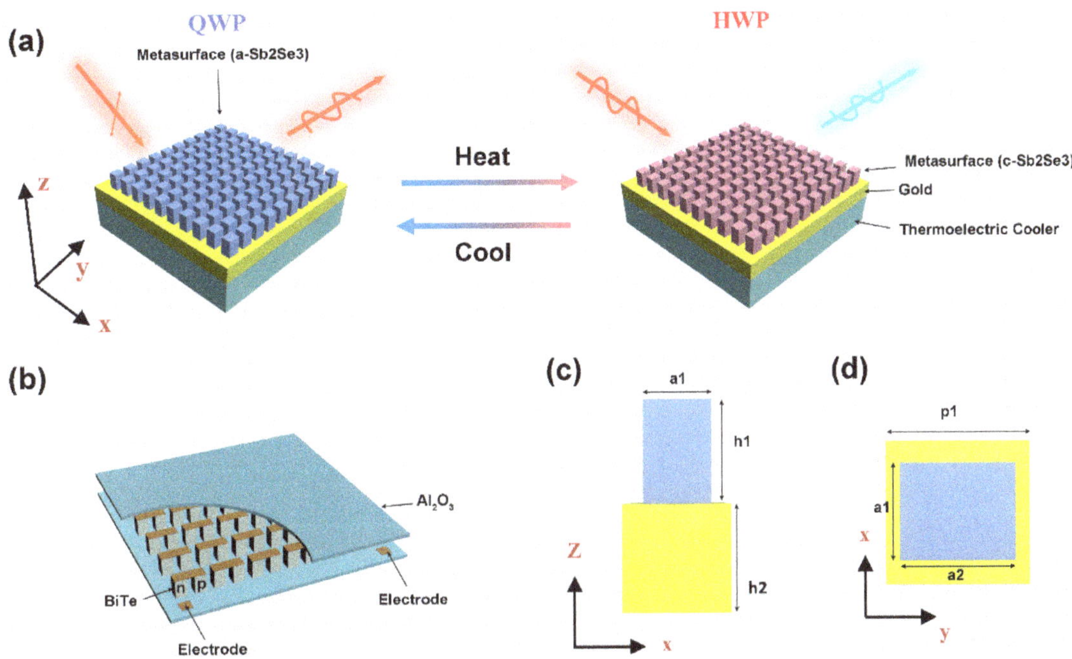

Figure 1. Phase-change metasurface. (**a**) Functional diagrams of the phase-change metasurface in different states. (**b**) Three-dimensional structure diagram of the TEC. (**c**) Front view of the metasurface unit cell. (**d**) Top view of the metasurface unit cell. $a1$ = 320 nm, $a2$ = 375 nm, $h1$ = 1200 nm, $p1$ = 580 nm, $h2$ = 600 nm. The gold layer is continuous, while the phase-change material Sb_2Se_3 layer has a periodic structure.

3. Results

Figure 2a presents the reflection spectra and phase-difference spectra of the amorphous phase-change metasurface within the 1.4 μm to 1.65 μm wavelength range. The light blue line corresponds to the reflection spectrum of the TM mode, featuring only one reflection valley, a, with typical asymmetry characteristics. With the exception of the redshift of the central wavelength corresponding to reflection valley b, the spectral profile of the reflection spectrum in the TE mode is analogous to that of the TM mode. Additionally, disregarding the effect of the reflection valleys, it can be discerned that the average reflection efficiency exceeds 85%, and the efficiencies of the TM and TE modes are proximate. The red line represents the phase difference, whose graph exhibits irregular oscillations. The purple region encompasses the phase space ranging from −100 degrees to −80 degrees. The phase-difference spectrum and the purple region have three overlapping locations. The first overlapping band contains reflection valley a, while the latter two overlapping bands lack reflection valleys. Consequently, in accordance with the definition of a quarter-waveplate, the phase-change metasurface can manifest the functionality of a reflective QWP in the latter two overlapping bands. Figure 2b displays the reflection spectra and phase-difference spectra of the metasurface in the crystalline state. The reflection spectrum of the TM mode has two reflection valleys, and the TE mode exhibits two reflection peaks. The average reflection rates of the TE and TM modes are similar, and the oscillation amplitudes are not significant. The phase-difference spectrum also has extensive overlap with the phase space from −170 degrees to −190 degrees, signifying that within the wavelength bands represented by these overlapping regions, the phase metasurface assumes the function of a half-waveplate.

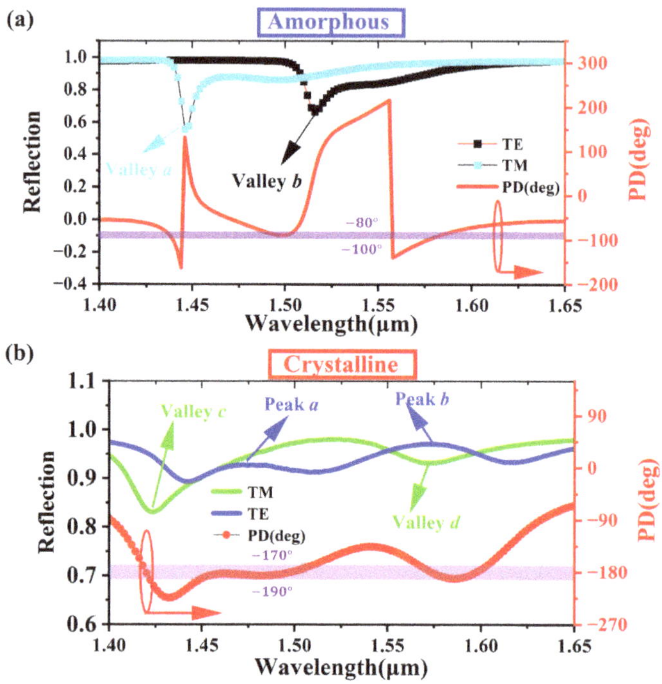

Figure 2. Reflection spectra and phase-difference spectra of the phase-change metasurface in two states. (**a**) a-Sb_2Se_3. (**b**) c-Sb_2Se_3. TM (TE) mode is defined as the case where the electric field direction of the linearly polarized incident light is along the X (Y) axis. The phase difference is the phase of the TM mode minus the phase of the TE mode.

In addition to employing the conventional reflection spectra and phase-difference spectra to evaluate the quality of waveplates, their polarization control characteristics can also be exploited to measure the function of waveplates. A quarter-waveplate possesses a phase delay of $\pi/2$ (an optical path difference of 1/4 wavelength), and thereby its main function is to convert linearly polarized light into circularly polarized light (or vice versa). Figure 3a depicts the polarization conversion function of the amorphous metasurface. Here, $LTC = \sqrt{(t_1 - n_1)^2 + (t_2 - n_2)^2 + (t_3 - n_3)^2 + (t_4 - n_4)^2}$, where (t_1, t_2, t_3, t_4) are the theoretical Stokes parameters of the reflected light, which are (1, 0, 0, 1) for RCP and (1, 0, 0, −1) for LCP when the incident light is 45° linearly polarized and −45° linearly polarized, respectively. (n_1, n_2, n_3, n_4) are the simulated Stokes parameters directly obtained from COMSOL 5.6. The smaller the value of LTC, the better the conversion from linearly polarized light to circularly polarized light. As presented in Figure 3a, the LTC spectra for the 45° linearly polarized light and −45° linearly polarized light are consistent. The values of LTC are less than 0.25 near the wavelengths of 1.5 μm and 1.58 μm, corresponding to the light blue rectangles. Figure 3b shows the polarization conversion function of the metasurface under the incidence of circularly polarized light. The definition of CTL here is the same as that in Figure 3a, but it quantitatively assesses the ability to convert circularly polarized light into linearly polarized light. The blue rectangles at the wavelengths of 1.5 μm and 1.58 μm also indicate that the CTL values within this band are less than 0.25. A half-waveplate (λ/2 waveplate) can introduce an additional π phase delay (an optical path difference of half a wavelength), which can alter the polarization direction of linearly polarized light or reverse the chirality (left-handed/right-handed) of circularly polarized light. Figure 3c,d present the cross-polarization and co-polarization efficiencies of the metasurface. Taking "Co-P(LCP)" and "Cross-P(LCP)" in the legend as examples, the LCP in parentheses indicates that the incident light on the metasurface is LCP. The value of Co-P(LCP) is the intensity of the LCP component in the reflected light that has the same polarization characteristics as the incident light (LCP) divided by the total intensity of the reflected light. The value of Cross-P(LCP) is the intensity of the RCP component in the reflected light that is orthogonal to the polarization of the incident light (LCP) divided by the total intensity of the reflected light. As shown in Figure 3c, for the linearly polarized incident light, the cross-polarization efficiency is greater than the co-polarization efficiency, and within the bandwidth represented by the red rectangle, the cross-polarization efficiency is greater than 90%. Figure 3d corresponds to the case of the circularly polarized incident light, presenting a similar situation as in Figure 3c, indicating that the metasurface possesses the characteristics of a half-waveplate. Additionally, the intersection of the light blue rectangle and the light red rectangle represents the actual bandwidth of the phase metasurface, which are the wavelengths of 1.485 μm to 1.505 μm and 1.57 μm to 1.59 μm, respectively.

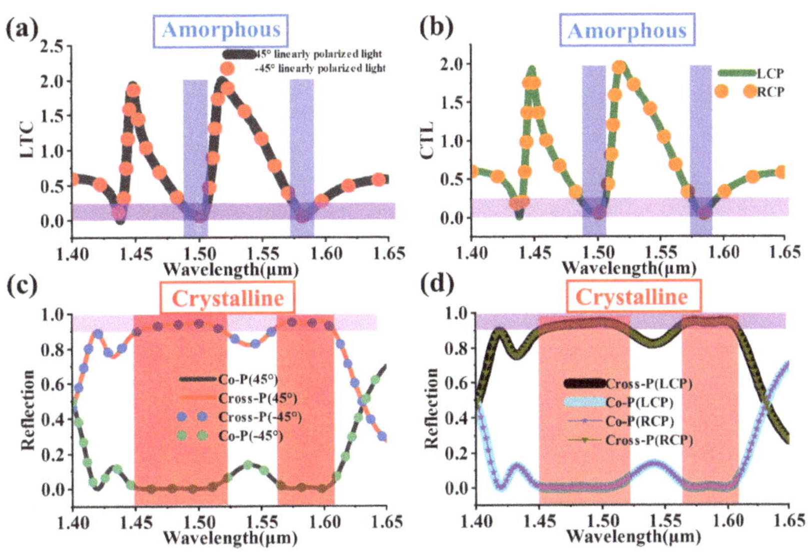

Figure 3. Polarization conversion capabilities of the phase-change metasurface in two states. (**a**) The ability to convert linearly polarized incident light into circularly polarized light. (**b**) The ability to convert circularly polarized incident light into linearly polarized light. (**c**) Cross-polarization efficiency and co-polarization efficiency in the case of linearly polarized incident light. (**d**) Cross-polarization efficiency and co-polarization efficiency in the case of circularly polarized incident light. The light blue rectangle and the light red rectangle respectively represent the effective bandwidth of the metasurface in the amorphous state and the crystalline state. The light purple area corresponds to the working range of the metasurface. The symbol "-45°" in the picture represents −45° polarized light.

3.1. Sb_2Se_3 in Low-Temperature (LT) (Amorphous) Phase

When the metasurface is in the amorphous state, the reflection spectra of the TM and TE modes present a typical asymmetry phenomenon. This asymmetric spectral line typically corresponds to Fano resonance, as depicted in Figure 4a. Fano resonance [34] is a quantum interference phenomenon induced by the interference between the discrete state (discrete energy level) and the continuous state (continuous energy level). Distinct from Lorentz (symmetric) resonance, Fano resonance exhibits a shape with one side being steep and the other side being gently sloped. The line shape of Fano resonance can be described by the Fano formula:

$$\sigma_{(E)} = \sigma_0 \frac{(q+\epsilon)^2}{1+\epsilon^2} + \sigma_b \tag{1}$$

$$\epsilon = \frac{E - E_r}{\Gamma/2} \tag{2}$$

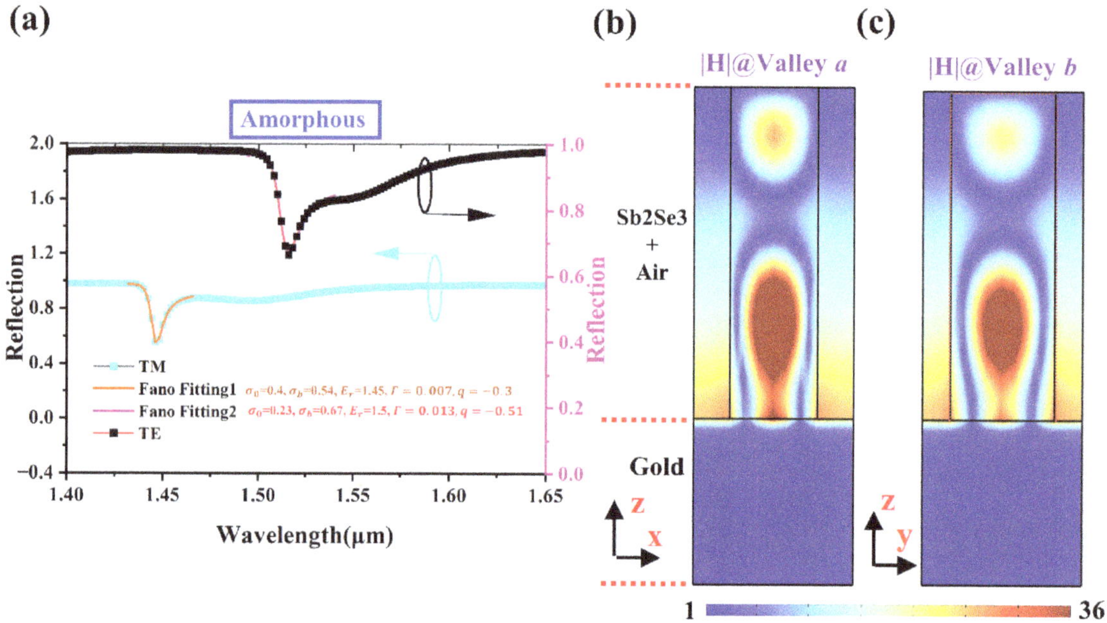

Figure 4. (**a**) Fitting of Fano resonance. The TE curve corresponds to the Y-axis on the right. (**b**,**c**) Magnetic field distribution of the metasurface corresponding to reflection valleys in the amorphous state of the phase-change material. The red dashed line is the boundary of the cubic block of phase-change material Sb_2Se_3.

Here, $\sigma_{(E)}$ represents the scattering or absorption cross-section, E_r is the resonance energy, σ_0 is the intensity scale of the discrete state's effect, Γ is the linewidth, q is the Fano parameter, σ_b represents the contribution of the non-resonant background, and E is the energy of the incident free photon. Through formula fitting, it can be discovered that when (σ_0, σ_b, E_r, Γ, q) = (0.4, 0.54, 1.45, 0.007, −0.3), the reflection spectrum of the TM mode can match Formulas (1) and (2). When (σ_0, σ_b, E_r, Γ, q) = (0.23, 0.67, 1.5, 0.013, −0.51), the reflection spectrum of the TE mode directly exported by the COMSOL software can be reconstructed by the Fano formula. The phenomenon that the reflection curve directly exported from the port of COMSOL perfectly coincides with Formulas (1) and (2) indicates that the optical resonance characteristics of the amorphous metasurface might be manipulated by Fano resonance. Figure 4b shows the XZ cross-sectional view of the magnetic field intensity of the phase-change metasurface in TM mode. There exists a typical magnetic field localization phenomenon within the phase-change material Sb_2Se_3, which might correspond to a special dielectric waveguide mode. The dielectric waveguide is composed of an air–Sb_2Se_3–air ternary planar waveguide, and its energy is confined within the high-refractive-index material, which is highly similar to the waveguide mode of the typical FP laser. Furthermore, at the boundary between the metal backplate and the phase-change material Sb_2Se_3, there seemingly exists a typical characteristic of the Surface Plasmon Polariton (SPP) [35] mode. This evanescent wave electric field can penetrate the metal and decays exponentially in the direction perpendicular to the interface. Figure 4c shows the YZ cross-sectional view of the magnetic field intensity of the phase-change metasurface in TE mode, and its resonance characteristics are similar to those in Figure 4b. Based on the above results, we conjecture that the spectral lines of the reflection valleys are generated by Fano resonance, and Fano resonance is caused by the interference between the dielectric waveguide mode and SPP.

3.2. *Sb_2Se_3 in High-Temperature (HT) (Crystalline) Phase*

Figure 5a,b depict the XZ cross-sectional magnetic field distribution maps of the crystalline metasurface at reflection valleys *c* and *d*. Multiple highly localized magnetic field spheres are arranged in a vertical row within the Sb_2Se_3 cubic block, with nearly no energy leakage into the air on both sides. This optical resonance with energy tightly confined in the lateral direction conforms to the characteristics of Fabry–Pérot (FP) resonance. This resonance causes the internal photons to oscillate repeatedly at the interface between the air and the phase-change material, accumulating a specific phase difference that matches the optical coherent constructive interference. Figure 5c presents the two-dimensional reflection spectrum associated with the geometric parameter $a1$ under the TM incident mode. The circles in Figure 5c correspond to the reflection valleys *c* and *d* in Figure 2b. It is evident that there exist reflection peak curves strongly correlated with the geometric parameter $a1$ which traverse the reflection valleys *c* and *d*. The significant redshift of the reflection resonance peaks with the increase in $a1$ is highly in accordance with the FP resonance. Figure 5d,e illustrate the electric field intensity distribution and the equivalent current direction diagrams of the metasurface at reflection peaks *a* and *b* under the TE mode. In contrast to Figure 5a,b, under the two reflection peaks, the electric field can leak from the phase-change material into the lateral air, and the electric field intensity diagrams do not correspond to significant optical resonance modes. Nevertheless, through observation, it can be found that the rotational directions of the equivalent currents within the phase-change material exhibit highly significant regularities. For peak *a*, two counterclockwise current loops and half a clockwise current loop appear successively from top to bottom. For peak *b*, two clockwise current loops and one counterclockwise current loop appear successively. We hypothesize that these internal current loops can be approximated as a series of electromagnetic radiation oscillation elements [36], and the electric field generated at the reflection end interferes with the TE incident mode, resulting in a reflection enhancement effect.

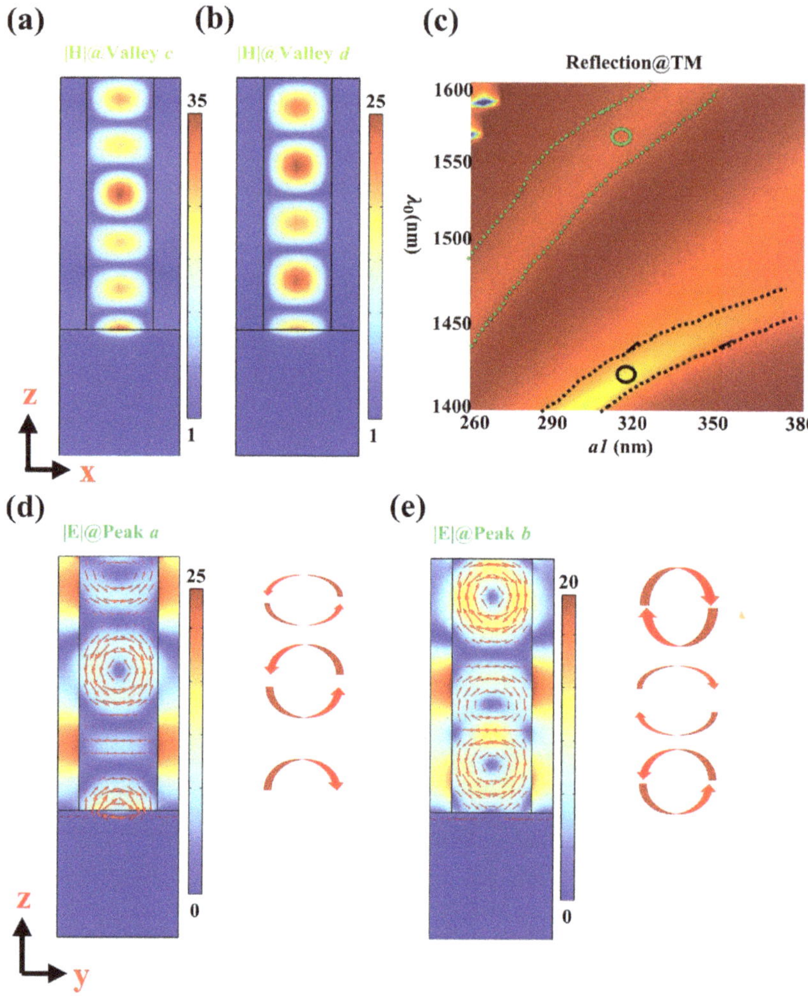

Figure 5. (**a**,**b**) Magnetic field distribution maps of the XZ cross-section of the phase-change material in the crystalline state, corresponding to the reflection valleys *c* and *d* of the TM mode of the metasurface. (**c**) TM mode reflection spectrum related to *a*1. The dashed line indicates the boundary of the reflection mode. (**d**,**e**) Electric field intensity distribution maps and equivalent current direction maps at the reflection peaks *a* and *b* of the TE mode, with their cross-sections being YZ. The red arrow on the right is an enlarged version of the equivalent current, serving a function of facilitating reading.

3.3. The Influence of Potential Process Errors on Metasurfaces

In the preceding chapters, we have presented the polarization regulation capabilities of the phase-change metasurface in two states and have also conducted a focused analysis of the corresponding optical resonance modes. In this chapter, we primarily discuss the impact of potential errors that may occur during the actual fabrication process of the metasurface on its performance. Figure 6a–d, respectively, display the images of the fillet radius *r*0, etching inclination angle θ, under-etching error $\Delta h1$, and over-etching error $\Delta h2$. Figure 6e–h, respectively, correspond to the LTC capabilities of the amorphous metasurface under the aforementioned errors. As depicted in Figure 6e, when the fillet radius *r*0 is less than 40 nm, it scarcely exerts any influence on the metasurface. Figure 6f indicates that as the etching inclination angle θ decreases, the LTC at wavelengths λ1 and λ2 initially increases and subsequently decreases. When θ is less than 88 degrees, the metasurface will gradually lose its function as a quarter-waveplate. Figure 6g reveals that when the under-etching error $\Delta h1$ does not exceed 40 nm, the metasurface can operate normally. Based on Figure 6h, it can be discerned that the performance loss of the metasurface under over-etching is more pronounced. When the over-etching error $\Delta h2$ is greater than 30 nm, the efficiency of LTC is reduced. Figure 6i–l, respectively, showcase the performance degradation of the crystalline metasurface under different processing errors, which can be evaluated by the cross-polarization efficiency. Figure 6i indicates that when the fillet radius $r0$ is less than 40 nm, the cross-polarization efficiency of the micro-device is greater than 0.9. Figure 6j shows that if the etching inclination angle θ can be greater than 88 degrees, the cross-polarization efficiency can also exceed 0.9. As shown in Figure 6k, when the under-etching error $\Delta h1$ is less than 40 nm, the performance of the metasurface is hardly compromised. Figure 6l indicates that the cross-polarization efficiency remains invariant initially and then declines as the over-etching error $\Delta h2$ increases.

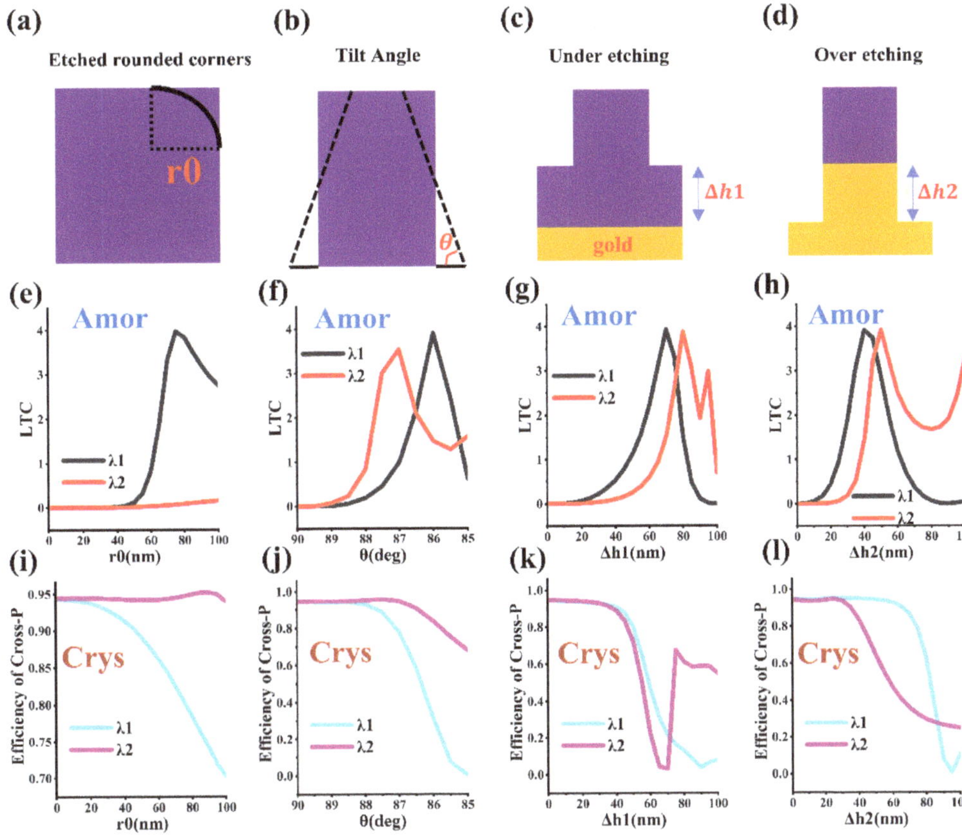

Figure 6. Analysis diagram of potential process errors based on COMSOL. (**a**–**d**) respectively correspond to the schematic diagrams of etched rounded corners, etched inclined angles, under-etching phenomena, and over-etching phenomena. The variables are $r0$, θ, $\Delta h1$, and $\Delta h2$ in sequence. λ1 and λ2 are 1.5 μm and 1.58 μm, respectively. The symbols "Amor" and "Crys", respectively, correspond to the amorphous and crystalline states of the phase-change metasurface. The variable corresponding to (**e**) and (**i**) is $r0$, the variable corresponding to (**f**) and (**j**) is θ, the variable corresponding to (**g**) and (**k**) is $\Delta h1$, and the variable corresponding to (**h**) and (**l**) is $\Delta h2$.

To validate the correctness of the COMSOL simulation model, we computed the impacts of several typical simulation variables on COMSOL. The discretization in the electromagnetic finite-element method transforms the continuous problem into a discrete algebraic system through polynomial shape functions, and the selection of the polynomial order and type directly influences the calculation accuracy. Figure 7a indicates that when the number of polynomial terms is greater than or equal to 2, the performance of the metasurface in both states scarcely changes. As depicted in Figure 7b, when the thickness of the PML layer ranges from 500 nm to 4000 nm, the calculation results of COMSOL remain consistent. To guarantee that no photons can penetrate the metal backplate, the thickness of the metal backplate is typically required to exceed its skin depth. Figure 7c shows that when the thickness of the metal backplate is greater than 120 nm, the effect of the metasurface becomes independent of the thickness of the metal backplate. Figure 7d indicates that when the mesh size of the phase-change material is less than 150 nm, the regulation effect of the metasurface has stabilized.

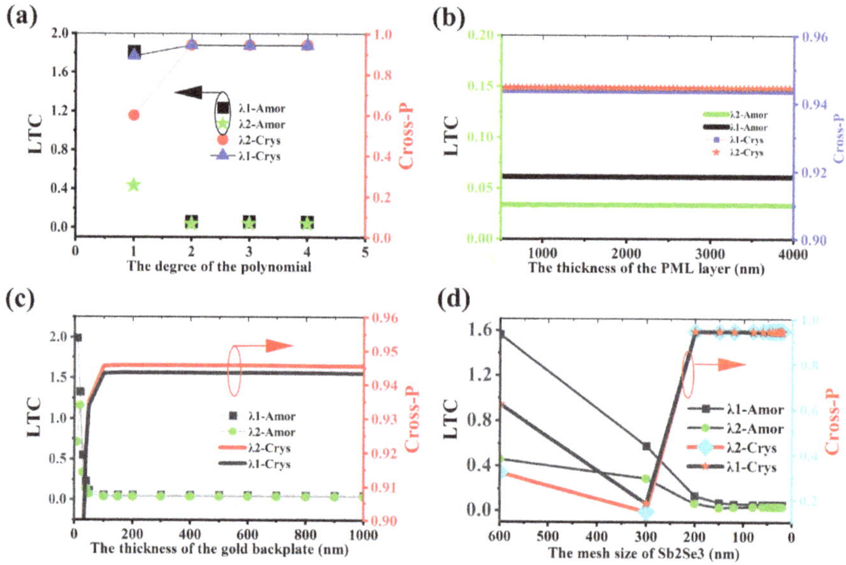

Figure 7. The influence of some typical simulation variables on the COMSOL model. (**a**) The type of numerical function for discretization. (**b**) The thickness of the PML layer. (**c**) The thickness of the metal backplate. (**d**) The maximum mesh size of the phase-change material Sb_2Se_3. The y-axis coordinate corresponding to the curve related to the red arrow is on the right side.

4. Discussion

We have designed a metasurface based on the phase-change material Sb_2Se_3, which can manifest the functionalities of a half-waveplate and a quarter-waveplate, respectively, under different conditions. A functional form has been achieved that is operable in multiple modes with only a single device. The phase-change metasurface consists of a TEC, a metal backplate, and a periodic Sb_2Se_3 graphic array from bottom to top. The temperature-controlled dual-function phase-change metasurface waveplate (HWP at high temperature/QWP at low temperature) we designed possesses revolutionary application potential, being capable of resolving multiple problems that traditional optical systems are unable to overcome, and is anticipated to pioneer new possibilities in dynamic optics, integrated photonics, and extreme environment optics.

Author Contributions: Conceptualization, B.C. and Y.Z.; methodology, Z.G.; software, B.C.; validation, B.C., Y.Z. and Z.G.; formal analysis, K.Z. and L.L.; investigation, K.Z. and L.L.; resources, G.S.; data curation, B.C.; writing—original draft preparation, B.C.; writing—review and editing, K.Z., G.S. and T.L.; visualization, Y.Z.; supervision, Z.G.; project administration, Z.G.; funding acquisition, G.S. All authors have read and agreed to the published version of the manuscript.

Funding: This research was funded by the Strategic Priority Research Program of the Chinese Academy of Sciences (Grant No. XDB0980000).

Data Availability Statement: The raw data supporting the conclusions of this article will be made available by the authors on request.

Acknowledgments: The authors are grateful to Jietao Liu of Nankai University and Zengxuan Jiang of UCAS.

Conflicts of Interest: The authors declare no conflicts of interest.

References

1. Goldstein, D.H. *Polarized Light*, 3rd ed.; CRC Press: Boca Raton, FL, USA, 2011. [CrossRef]
2. Zhang, L.L.; Zhong, H.; Deng, C.; Zhang, C.L.; Zhao, Y.J. Polarization sensitive terahertz time-domain spectroscopy for birefringent materials. *Appl. Phys. Lett.* **2009**, *94*, 904. [CrossRef]
3. Wu, Q.H.; Hodgkinson, I.J. Materials for Birefringent Coatings. *Appl. Opt.* **1994**, *33*, 8109–8110. [CrossRef] [PubMed]
4. Chen, C.; Gao, S.; Xiao, X.; Ye, X.; Wu, S.; Song, W.; Li, H.; Zhu, S.; Li, T. Highly Efficient Metasurface Quarter-Wave Plate with Wave Front Engineering. *Adv. Photonics Res.* **2021**, *2*, 2000154. [CrossRef]
5. Ding, F.; Wang, Z.; He, S.; Shalaev, V.M.; Kildishev, A.V. Broadband High-Efficiency Half-Wave Plate: A Supercell-Based Plasmonic Metasurface Approach. *ACS Nano* **2015**, *9*, 4111–4119. [CrossRef]
6. Turunen, J.; Friesem, A.A.; Friberg, A.T. Polarization Optics. *J. Opt. A Pure Appl. Opt.* **2011**, *6*, S1. [CrossRef]
7. Meng, W.; Fröch, J.E.; Cheng, K.; Pi, D.; Li, B.; Majumdar, A.; Maier, S.A.; Ren, H.; Gu, M.; Fang, X. Ultranarrow-linewidth wavelength-vortex metasurface holography. *Sci. Adv.* **2025**, *11*, eadt9159. [CrossRef]
8. Kuznetsov, A.I.; Brongersma, M.L.; Yao, J.; Chen, M.K.; Levy, U.; Tsai, D.P.; Zheludev, N.I.; Faraon, A.; Arbabi, A.; Yu, N.; et al. Roadmap for Optical Metasurfaces. *ACS Photonics* **2024**, *11*, 816–865. [CrossRef]
9. Yang, Y.; Lee, E.; Park, Y.; Seong, J.; Kim, H.; Kang, H.; Kang, D.; Han, D.; Rho, J. The Road to Commercializing Optical Metasurfaces: Current Challenges and Future Directions. *ACS Nano* **2025**, *19*, 3008–3018. [CrossRef]
10. Kowerdziej, R.; Wróbel, J.; Kula, P. Ultrafast electrical switching of nanostructured metadevice with dual-frequency liquid crystal. *Sci. Rep.* **2019**, *9*, 20367. [CrossRef]
11. Bhowmik, T.; Kr. Bhowmik, B.; Kr. Pandey, P.; Kumar, G.; Sikdar, D. Dual-band electro-optic modulator based on tunable broadband metamaterial absorber. *Opt. Laser Technol.* **2023**, *161*, 109129. [CrossRef]
12. Ma, T.; Zhao, K.; Gu, M.; Zhou, H.; Liu, C.; Cheng, C.; Dong, Q.; Ma, L. Longitudinal Multi-Channel Focused Vortex and Vector Beams Generation by Quarter-Wave Plate Meta-Atom Metasurfaces. *Nanomaterials* **2025**, *15*, 324. [CrossRef] [PubMed]
13. Deng, Y.; Wu, C.; Meng, C.; Bozhevolnyi, S.I.; Ding, F. Functional Metasurface Quarter-Wave Plates for Simultaneous Polarization Conversion and Beam Steering. *ACS Nano* **2021**, *15*, 18532–18540. [CrossRef]
14. He, H.; Tang, S.; Zheng, Z.; Ding, F. Multifunctional all-dielectric metasurface quarter-wave plates for polarization conversion and wavefront shaping. *Opt. Lett.* **2022**, *47*, 2478–2481. [CrossRef]
15. Li, T.; Hu, X.; Chen, H.; Zhao, C.; Xu, Y.; Wei, X.; Song, G. Metallic metasurfaces for high efficient polarization conversion control in transmission mode. *Opt. Express* **2017**, *25*, 23597–23604. [CrossRef] [PubMed]
16. Li, H.; Li, J.; Zhao, C.; Zheng, C.; Xu, H.; Xu, W.; Tan, Q.; Song, C.; Shen, Y.; Yao, J. Meta-optics empowered by all-silicon quarter-wave plates for generating focused vortex beams. *Chin. J. Phys.* **2024**, *92*, 651–663. [CrossRef]
17. Zhao, Y.; Alù, A. Tailoring the dispersion of plasmonic nanorods to realize broadband optical meta-waveplates. *Nano Lett.* **2013**, *13*, 1086–1091. [CrossRef] [PubMed]
18. Yu, N.; Aieta, F.; Genevet, P.; Kats, M.A.; Gaburro, Z.; Capasso, F. A Broadband, Background-Free Quarter-Wave Plate Based on Plasmonic Metasurfaces. *Nano Lett.* **2012**, *12*, 6328–6333. [CrossRef]
19. Kebci, Z.; Hamidi, M.; Zeghdoudi, T.; Belkhir, A.; Lamrous, O.; Baida, F.I. Design of a metallic half-wave plate based on birefringent metamaterial for applications in the visible range. *Phys. Scr.* **2025**, *100*, 035552. [CrossRef]
20. Li, Z.; Liu, W.; Cheng, H.; Chen, S.; Tian, J. Realizing Broadband and Invertible Linear-to-circular Polarization Converter with Ultrathin Single-layer Metasurface. *Sci. Rep.* **2015**, *5*, 18106. [CrossRef]

21. Huang, X.; Ma, X.; Gao, H.; Guo, L.; Li, X. Ultra-wideband linear-polarization conversion metasurface with high-efficient asymmetric transmission. *Appl. Phys. A* **2023**, *129*, 278. [CrossRef]
22. Michel, A.-K.U.; Meyer, S.; Essing, N.; Lassaline, N.; Lightner, C.R.; Bisig, S.; Norris, D.J.; Chigrin, D.N. The Potential of Combining Thermal Scanning Probes and Phase-Change Materials for Tunable Metasurfaces. *Adv. Opt. Mater.* **2021**, *9*, 2001243. [CrossRef]
23. Landau, L. The Theory of Phase Transitions. *Nature* **1936**, *138*, 840–841. [CrossRef]
24. Xiao, S.; Chettiar, U.K.; Kildishev, A.V.; Drachev, V.; Khoo, I.C.; Shalaev, V.M. Tunable magnetic response of metamaterials. *Appl. Phys. Lett.* **2009**, *95*, 033115. [CrossRef]
25. Shcherbakov, M.R.; Vabishchevich, P.P.; Shorokhov, A.S.; Chong, K.E.; Choi, D.-Y.; Staude, I.; Miroshnichenko, A.E.; Neshev, D.N.; Fedyanin, A.A.; Kivshar, Y.S. Ultrafast All-Optical Switching with Magnetic Resonances in Nonlinear Dielectric Nanostructures. *Nano Lett.* **2015**, *15*, 6985–6990. [CrossRef] [PubMed]
26. Kafaie Shirmanesh, G.; Sokhoyan, R.; Pala, R.A.; Atwater, H.A. Dual-Gated Active Metasurface at 1550 nm with Wide (>300°) Phase Tunability. *Nano Lett.* **2018**, *18*, 2957–2963. [CrossRef]
27. Chen, G.Y.; Dneg, B.; Cai, G.B.; Zhang, T.K.; Dong, W.F.; Zhang, W.X.; Xu, A.W. The Fractal Splitting Growth of Sb_2S_3 and Sb_2Se_3 Hierarchical Nanostructures. *J. Phys. Chem. C* **2008**, *112*, 672–679. [CrossRef]
28. Zhou, Y.; Wang, L.; Chen, S.; Qin, S.; Liu, X.; Chen, J.; Xue, D.J.; Luo, M.; Cao, Y.; Cheng, Y. Thin-film Sb_2Se_3 photovoltaics with oriented one-dimensional ribbons and benign grain boundaries. *Nat. Photonics* **2015**, *9*, 409–415. [CrossRef]
29. Chein, R.; Huang, G. Thermoelectric cooler application in electronic cooling. *Appl. Therm. Eng.* **2004**, *24*, 2207–2217. [CrossRef]
30. Zou, H.; Rowe, D.M.; Williams, S.G.K. Peltier effect in a co-evaporated Sb_2Te_3(P)-Bi_2Te_3(N) thin film thermocouple. *Thin Solid Film.* **2002**, *408*, 270–274. [CrossRef]
31. Drebushchak, V.A. The Peltier effect. *J. Therm. Anal. Calorim.* **2008**, *91*, 311–315. [CrossRef]
32. Delaney, M.; Zeimpekis, I.; Lawson, D.; Hewak, D.W.; Muskens, O.L. A New Family of Ultralow Loss Reversible Phase-Change Materials for Photonic Integrated Circuits: Sb_2S_3 and Sb_2Se_3. *Adv. Funct. Mater.* **2020**, *30*, 2002447. [CrossRef]
33. Babar, S.; Weaver, J.H. Optical constants of Cu, Ag, and Au revisited. *Appl. Opt.* **2015**, *54*, 477–481. [CrossRef]
34. Luk'Yanchuk, B.; Zheludev, N.I.; Maier, S.A.; Halas, N.J.; Nordlander, P.; Giessen, H.; Chong, C.T. The Fano resonance in plasmonic nanostructures and metamaterials. *Nat. Mater.* **2010**, *9*, 707. [CrossRef] [PubMed]
35. Bozhevolnyi, S.I.; Erland, J.; Leosson, K.; Skovgaard, P.M.W.; Hvam, J.M. Waveguiding in Surface Plasmon Polariton Band Gap Structures. *Phys. Rev. Lett.* **2001**, *86*, 3008–3011. [CrossRef]
36. Gansel, J.K.; Thiel, M.; Rill, M.S.; Decker, M.; Bade, K.; Saile, V.; von Freymann, G.; Linden, S.; Wegener, M. Gold Helix Photonic Metamaterial as Broadband Circular Polarizer. *Science* **2009**, *325*, 1513–1515. [CrossRef]

crystals

MDPI

Article

Evolution of Phonon Spectral Energy Density in Superlattice Structures

Milad Nasiri * and Yan Wang *

Department of Mechanical Engineering, University of Nevada, Reno, Reno, NV 89557, USA
* Correspondence: mnasiri@unr.edu (M.N.); yanwang@unr.edu (Y.W.)

Abstract: Superlattices are a distinctive class of artificial nanostructures formed by the periodic stacking of two or more materials. The high density of interfaces in these structures often gives rise to exotic physical properties. In the context of thermal transport, it is well established that such interfaces can significantly scatter particle-like phonons while also inducing constructive or destructive interference in wave-like phonons, depending on the relationship between the phonons' coherence lengths and the superlattice's period thickness. In this work, we systematically investigate the effect of temperature on the spectral energy density of phonon modes in superlattices. Additionally, we examine how variations in superlattice period thickness influence phonon lifetimes and energy density. Our findings provide critical insights into the spectral phonon properties of superlattices, particularly in terms of their coherence and lifetimes.

Keywords: superlattice; phonon; coherence; spectral energy density; lifetime; group velocity; thermal conductivity

1. Introduction

Superlattices are a class of artificially engineered metamaterials that enable precise control over phonon band structures, resulting in thermal and acoustic properties that deviate significantly from those of their constituent materials. These engineered characteristics make superlattices highly attractive for thermoelectric applications, where their ability to substantially suppress lattice thermal conductivity via interfacial phonon scattering enhances energy conversion efficiency [1–4]. In addition, superlattices are integral components in nanophotonic devices such as quantum cascade lasers, where they critically influence optical and electronic performance. However, real-world imperfections—such as variations in layer thickness, interfacial defects, and complex thermodynamic environments—introduce intricate thermal transport phenomena. These complexities necessitate a deeper understanding of phonon dynamics in order to optimize heat dissipation and improve the thermal performance of superlattice-based devices.

Phonon transport in superlattices can be broadly categorized into two regimes: coherent and incoherent transport [5–9]. In the coherent regime, phonons maintain their wave-like nature across multiple interfaces, leading to transport behaviors that deviate from classical particle-like predictions such as those described by the Boltzmann transport equation. Intriguingly, experimental and simulation studies have shown that reducing the superlattice period length can paradoxically increase thermal conductivity [10–13], while increasing the number of superlattice periods can result in a roughly linear rise in thermal conductivity [5,6,14,15]. These observations challenge the classical view, in which denser interfaces should increase phonon interface scattering and suppress thermal conductivity, which should be saturated once the phonon mean free paths are limited by interface density.

When phonon coherence is significant, the superlattice behaves as a periodic medium whose unit cell corresponds to one superlattice period [16,17]. In this case, the group velocity of coherent phonon modes can increase as the period length decreases, leading to a rise in thermal conductivity—a trend opposite to that expected under incoherent transport. This non-monotonic dependence of thermal conductivity on layer thickness has been observed across a variety of material systems, both experimentally and numerically [10–13]. Additionally, short-period superlattices often exhibit a reduced anharmonic phonon scattering phase space, which further contributes to the increase in thermal conductivity. Despite these well-established indicators of phonon coherence, most studies to date have primarily focused on macroscopic trends in thermal conductivity. A more fundamental understanding of coherence requires mode-resolved investigations of phonon behavior under varying thermal and structural conditions.

Despite decades of intensive research on superlattice structures [7,8,18,19], since the 1990s in particular, the study of coherent phonon transport and its manipulation in superlattices has emerged as a rapidly growing area of interest. Since 2014, several landmark experimental studies have uncovered distinct and significant coherent phonon behaviors. Notable examples include the seminal work by Ravichandran et al. (2014), which demonstrated a crossover from coherent to incoherent phonon transport in $(SrTiO_3)_m/(CaTiO_3)_n$ superlattices [12]; the study by Luckyanova et al. (2018) on the Anderson localization of coherent phonons in GaAs/AlAs superlattices [20]; and experiments performed by Hu et al. (2020), which showed that aperiodic GaAs/AlAs superlattices exhibit lower thermal conductivity than their periodic counterparts.

Recent advances in computational and experimental methodologies have enabled more nuanced investigations into the transition between coherent and incoherent phonon transport in superlattices. Molecular dynamics simulations have clarified how key parameters—including temperature [15], superlattice period length [6,11,13], the number of periods [21], interfacial defects [22,23], and aperiodicity in layer thicknesses [6,16,24,25]—govern this transition. Anharmonic phonon scattering, in particular, has been identified as a key mechanism that disrupts coherence by breaking the phase relationships between phonon modes [15,17]. Wavepacket simulations further demonstrate that phonons entering a superlattice from a homogeneous medium can form new, spatially extended coherent modes governed by modified phonon dispersion relations. These modes traverse the superlattice with minimal interface scattering, effectively perceiving the entire structure as a homogeneous medium [10,12,16,25].

Such coherent phonon modes provide new opportunities for researchers to tailor thermal conductivity through structural modifications. Strategies such as introducing random variations in layer thickness [6,24,26] or embedding nanoscale obstacles [20] have been shown to induce phonon localization and reduce thermal conductivity. Notably, our recent work demonstrated that by optimizing the thicknesses of Lennard-Jones superlattice layers and incorporating alloying, it is possible to achieve thermal conductivities even lower than those of the binary alloys formed by the base materials [22]. Experimental results from Si/Ge superlattices demonstrate similar effects [27].

A particularly informative approach to examining the coherent–incoherent transition involves analyzing phonon behavior in reciprocal space. Wavepacket simulations show that as phonons traverse into a superlattice, their spectral content evolves, forming superlattice-specific coherent modes [16,25]. However, most such simulations assume a background temperature of zero, raising questions about how these coherent modes persist under realistic thermal conditions. To address this, recent analyses of local heat flux spectra in conceptual "m40–m90" superlattices have provided direct evidence of phonon coherence and its degradation with increasing temperature or period length [17]. In short-period superlattices at low temperatures, the nearly identical spectra across both material regions indicate strong coherence. As the temperature or period length increases, these spectra diverge, signaling the emergence of incoherent transport.

In our recent work [15], we employed nonequilibrium molecular dynamics simulations to study periodic and aperiodic Lennard-Jones superlattices. By fitting the results to a two-mode coherent–incoherent phonon model, we uncovered a significant disappearance of coherent phonon modes, i.e., decoherence. This effect was attributed to the conversion of coherent modes to incoherent modes, driven by phase-breaking anharmonic scattering processes that become increasingly dominant at elevated temperatures.

Despite significant progress in the field, a comprehensive understanding of how key phonon properties—particularly the population and lifetimes of coherent phonon modes—evolve with temperature and structural parameters remains incomplete. Most previous investigations [28,29] have concentrated on low-temperature regimes and short superlattice periods, where phonon coherence dominates. In this study, we extend the scope to higher temperatures and longer period lengths to capture the full transition from predominantly coherent to incoherent phonon transport. Through this work, we aim to uncover the fundamental mechanisms that govern heat conduction in superlattices and provide guidance for the rational design of thermally optimized superlattice structures.

This paper is organized as follows: Section 2 outlines the computational methodologies used in this study. Section 3 presents and discusses our results, focusing on the evolution of spectral phonon properties as a function of temperature and superlattice period length. Finally, Section 4 summarizes our key findings and their implications for phonon transport engineering in superlattices.

2. Methodology

2.1. Model System

Molecular dynamics (MD) simulations were performed using the LAMMPS package [30], which is a widely used classical molecular dynamics code, to investigate phonon behaviors in model superlattice structures. A face-centered cubic (FCC) superlattice crystal was used as the model system, and atomic interactions were described using the Lennard-Jones (LJ) potential:

$$\Phi(r_{ij}) = 4\epsilon\left[\left(\frac{\sigma}{r_{ij}}\right)^{12} - \left(\frac{\sigma}{r_{ij}}\right)^{6}\right], \tag{1}$$

where $\Phi(r_{ij})$ denotes the pairwise interaction energy between atoms i and j, r_{ij} is the interatomic distance, σ is the zero-potential distance, and ϵ represents the depth of the potential well. Table 1 displays all the parameters of the LJ potential and atomic mass that were adopted in our simulations. Specifically, the parameters $\sigma = 0.34$ nm and $\epsilon = 0.1664$ eV were adopted. The value of ϵ was chosen to be 16 times greater than that of solid argon in order to emulate the stronger atomic bonding typical of covalently bonded semiconductors, such as Si/Ge [31,32], GaAs/AlAs [5,20], and Bi_2Te_3/Sb_2Te_3 [1] superlattices. The LJ potential offers a computationally efficient means of capturing essential interatomic interactions while avoiding the higher cost of more complex potentials such as the Tersoff potential, thereby allowing a clearer analysis of underlying physical mechanisms.

To explore the effects of both temperature and structure on phonon properties, superlattice configurations with varying period lengths were considered: 1UC-1UC, 2UC-2UC, 3UC-3UC, 4UC-4UC, and 8UC-8UC. Their cross-sectional dimensions were fixed at 4UC × 4UC in the *y*–*z* plane, where the conventional unit cell (UC) of the FCC lattice was taken to be 5.34 Å. Both constituent layers of the superlattices shared identical LJ parameters, implying equal lattice constants and bonding strengths. The only distinction between the two materials was their atomic mass: one had a mass of 40 g/mol (analogous to silicon, denoted as `m40`) and the other had a mass of 90 g/mol (analogous to germanium, denoted as `m90`). A schematic illustration of representative superlattice structures is shown in Figure 1.

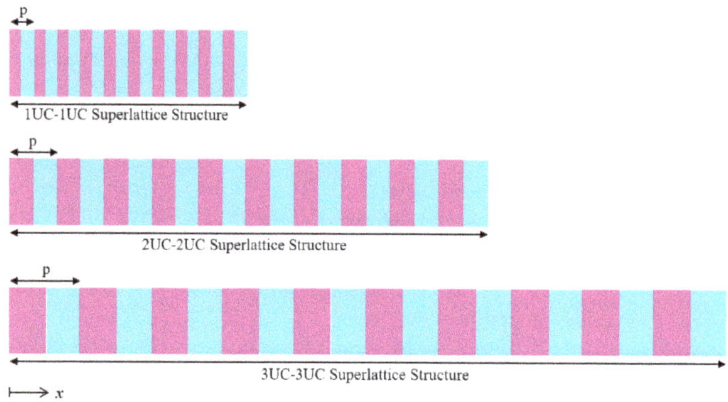

Figure 1. Schematic representations of the 1UC-1UC, 2UC-2UC, and 3UC-3UC superlattice structures. Purple and cyan denote `m40` and `m90` atoms, respectively.

Table 1. Material parameters used in the model.

Zero-Energy Crossing Distance	Potential Well Depth	Cutoff Radius	Atomic Mass of 1st Atom Type	Atomic Mass of 2nd Atom Type
σ	ε	r	m40 mass	m90 mass
0.34 nm	0.1664 eV	10 Å	40 g/mol	90 g/mol

2.2. Equilibrium Molecular Dynamics Simulations

Periodic boundary conditions were imposed in all three spatial directions to eliminate phonon scattering from the simulation domain boundaries, thereby emulating a bulk-like environment with infinite periodicity in the transport direction. Consequently, phonons encountered interface scattering but were free from artificial reflections at the simulation box's edges, allowing us to effectively mimic an extended superlattice system.

The simulations were initiated by assigning phonons random atomic velocities sampled from a Gaussian distribution corresponding to an average system temperature of 5 K. Two-stage isothermal–isobaric (NPT) ensemble simulations were then employed for structural equilibration. In the first stage, the temperature was ramped from 5 K to the target temperature over 2 ns (2 million time steps with a 1 fs time step). The second stage kept the system at the desired temperature for an additional 4 ns to ensure full relaxation and equilibration.

After the NPT stages, the simulation transitioned to the canonical (NVT) ensemble, during which atomic velocity trajectories were recorded. These trajectories served as the basis for a spectral energy density analysis, which provides frequency- and wavevector-resolved information on phonon properties such as their lifetime and dispersion. Details of the spectral energy density methodology and its application in this study are provided in the next subsection and discussed further in Section 3.

2.3. Phonon Spectral Energy Density Analysis

A phonon spectral energy density (SED) analysis was used to extract detailed phonon lifetimes and dispersion relations from the equilibrium MD simulations. The formalism of this SED analysis involves projecting the atomic displacements onto the crystal's normal mode coordinates in the time domain:

$$q_{k,\nu}(t) = \sum_{\alpha=1}^{3} \sum_{b=1}^{n} \sum_{l=1}^{N_c} \sqrt{\frac{m_b}{N_c}}\, u_\alpha^{l,b}(t)\, e_{b,\alpha}^{k,\nu*}\, e^{ik\cdot r_0^l}, \quad (2)$$

where $u_\alpha^{l,b}(t)$ denotes the α-component of the time-dependent displacement of atom b in unit cell l, N_c is the number of unit cells, m_b is the atomic mass, $e_{b,\alpha}^{k,\nu*}$ is the complex conjugate of the phonon eigenvector, and r_0^l is the equilibrium position of unit cell l. The mode index ν indicates the phonon branch, and k is the phonon wavevector.

The spectral energy density function $\Phi_{k,\nu}(\omega)$ is obtained via the Fourier transform of the time derivative of $q_{k,\nu}(t)$:

$$\Phi_{k,\nu}(\omega) = \left|\mathcal{F}[\dot{q}_{k,\nu}(t)]\right|^2 = \frac{C_{k,\nu}}{(\omega - \omega_{k,\nu}^A)^2 + (\tau_{k,\nu}^{-1})^2/4}, \quad (3)$$

where $\omega_{k,\nu}^A$ is the anharmonic frequency of the phonon mode, $\tau_{k,\nu}^{-1}$ is the phonon scattering rate (inverse lifetime), and $C_{k,\nu}$ is a mode-specific constant. The Lorentzian form enables the direct extraction of phonon lifetimes and frequencies by curve-fitting the computed SED.

While the SED approach enables the extraction of phonon dispersion at finite temperatures—facilitating the investigation of phonon properties in complex materials [33]—we also performed harmonic lattice dynamics calculations using the ALAMODE package [34]. ALAMODE is a widely used lattice dynamics software capable of computing phonon properties in conjunction with molecular dynamics packages (e.g., LAMMPS) or density functional theory codes. These calculations were used to obtain phonon dispersion curves for various superlattice configurations, providing complementary insights and serving as a means to validate our SED analysis.

2.4. Convergence Test

To assess the impact of the trajectory length on SED accuracy, a convergence study was conducted by computing SEDs using different numbers of MD steps. Figure 2 presents the SED contours for a 2UC-2UC superlattice comprising 32 bilayers, which was calculated at 50 K using 1.5, 5, 10, and 20 million MD steps.

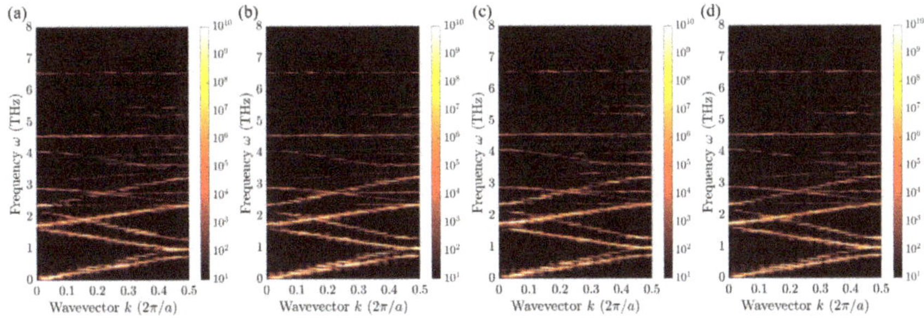

Figure 2. Convergence of the SED with respect to the number of MD steps used. The SED is computed for a 2UC-2UC superlattice with 32 bilayers at 50 K using (**a**) 1.5 million, (**b**) 5 million, (**c**) 10 million, and (**d**) 20 million MD steps. Also, "a" is the period length of the superlattice.

As seen in Figure 2, the overall features and resolution of the SED remain consistent even when 1.5 million MD steps are used. No significant improvement is observed beyond 5 million steps, indicating that this dataset provides a favorable balance between accuracy and computational efficiency. Therefore, all subsequent analyses were conducted using 5 million MD steps for SED calculations.

3. Results and Discussion

In this section, we present key results from the phonon SED analysis of Lennard-Jones superlattices across a broad range of temperatures and period lengths. Both factors have been identified in previous studies as critical in determining the phonon coherence and thermal transport behavior of superlattice structures, but it remains largely unexplored how individual coherent phonon modes, i.e., the superlattice modes, are affected by these factors.

3.1. Temperature Dependence of SED Heat Map

Figure 3 presents the SED heat map of a 32-period 2UC-2UC superlattice across a temperature range of 10 K to 150 K. In all cases, the SED heat map reveals distinct acoustic branches alongside multiple optical branches. When comparing the low-temperature cases (e.g., 10 K and 20 K) with high-temperature cases (e.g., 140 K and 150 K), we observe that the increasing temperature enhances the density of the energy carried by each phonon mode. This behavior aligns with phonon theory, which predicts greater energy storage in phonon modes at elevated temperatures.

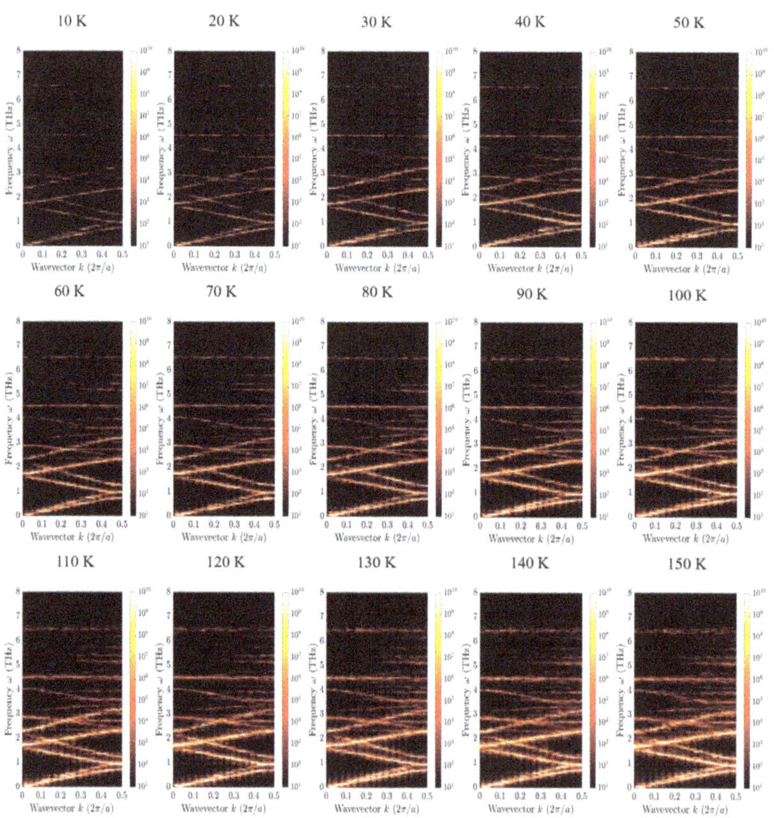

Figure 3. The phonon spectral energy density heat maps of a 32-period 2UC-2UC Lennard-Jones conceptual superlattice, obtained from equilibrium molecular dynamics simulations at different temperatures.

Furthermore, the SED heat maps qualitatively illustrate the significant broadening of prominent phonon branches—particularly low-lying acoustic modes and intermediate-frequency optical modes—both in the frequency and wavevector domains. The frequency-domain broadening of phonon linewidths indicates a reduction in phonon lifetime, primarily due to intensified anharmonic phonon–phonon scattering at higher temperatures. However, the slopes of the phonon branches, which correspond to phonon group velocities, remain largely unchanged with temperature. Consequently, the phonon mean free path, given by the product of the phonon group velocity and phonon lifetime, decreases as the temperature rises. These trends are consistent with previous studies on simpler material systems [35,36].

Additionally, the observed broadening along the wavevector axis suggests a reduction in phonon coherence length [37,38]. While the phonon mean free path is frequently discussed in thermal transport studies, the phonon coherence length is less explored, yet it plays a crucial role in determining phonon coherence in superlattice structures [33,39].

It is important to note that our classical molecular dynamics simulations correspond to the classical limit of phonon distribution, where phonon occupation follows the Maxwell–Boltzmann distribution rather than the Bose–Einstein distribution. As a result, even at 10 K, we still observe significant optical phonon branches extending to high frequencies near 6.5 THz. Nevertheless, the SED analysis presented here provides key insights into the temperature-dependent variations in phonon lifetime and population. A more detailed examination of these effects, along with their structural influences, will be discussed later in this manuscript.

3.2. Spectral Energy Density as a Function of Superlattice Period Length

The influence of the period length p on the phonon properties of superlattices has been extensively studied [1,10–13]. The general consensus is that when thermal transport is dominated by incoherent, particle-like phonons, thermal conductivity increases with the increasing period length due to reduced interface density and, consequently, decreased phonon interface scattering. Conversely, when coherent phonons dominate, thermal conductivity decreases as the period length increases. This behavior arises because the phonon group velocity typically decreases in superlattices with longer periods (or unit cells), as explained by the Brillouin zone folding mechanism [10], which was proposed in the 2000s to describe period-dependent lattice thermal transport in superlattices. More recently, it has also been shown that a longer period length is associated with an increased phonon scattering rate due to changes in the phonon scattering phase space [29]. This phenomenon is analogous to that observed in simpler materials, such as diamond and boron nitride [40], and in nanomesh structures (also known as phononic crystals) [33], where a larger unit cell complicates the phonon band structure, significantly increasing the number of three-phonon scattering processes allowed.

Here, we examine the evolution of phonon SED as a function of the period length. As shown in Figure 4, the upper row presents the phonon dispersion relations of m40-m90 superlattices with varying period lengths, ranging from the smallest possible period of 1UC-1UC (composed of one unit cell of m40 material and one unit cell of m90 material) to an 8UC-8UC superlattice, which represents the upper limit of the unit cell size that can be handled by our computational resources for lattice dynamics calculations. The width of the horizontal axis in each panel corresponds to the size of the first Brillouin zone, meaning that superlattices with longer period lengths in real space exhibit smaller first Brillouin zones, resulting in narrower phonon dispersion plots.

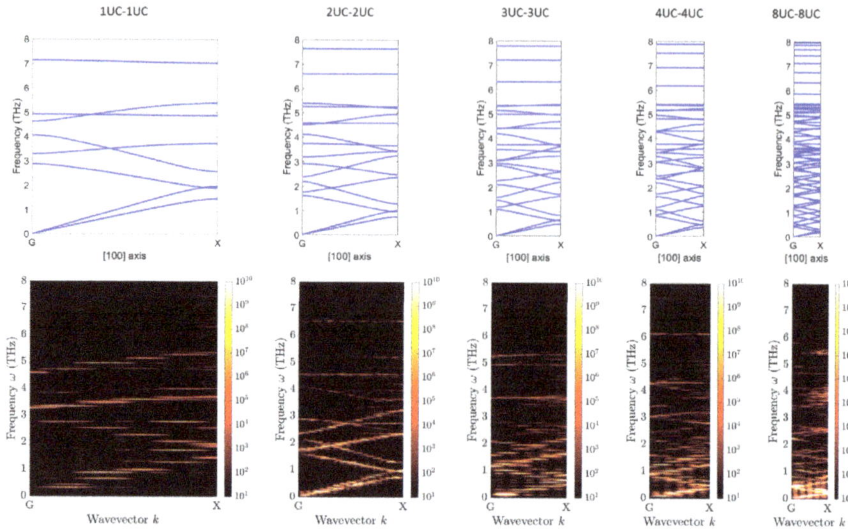

Figure 4. Optical- and acoustic-mode frequency shifting over different periods and the phonon spectral energy density of superlattices of different period lengths at 50 K.

A comparison of the five phonon dispersion panels in Figure 4 shows that although the overall structure of the phonon branches remains similar, the slopes of the acoustic branches systematically decrease with the increasing period length. This reduction in slope is most pronounced near the Brillouin zone boundary (close to the X point), where the acoustic branches progressively flatten, approaching a slope of zero. This reduction in slope indicates a decrease in phonon group velocity, a behavior well understood within the framework of Brillouin zone folding [10]. The SED heat maps in the lower row of Figure 4 confirm this trend, as they exhibit phonon dispersion curves that are in good agreement with the lattice dynamics calculations presented in the upper row. These heat maps also show a significant flattening of the acoustic branches. Furthermore, the phonon branches appear increasingly broadened along the vertical (frequency) axis, indicating an increased phonon scattering rate. This observation aligns with previous studies, which suggest that increased phonon scattering at longer period lengths contributes to changes in thermal transport behavior. In the following sections, we will provide a more detailed analysis of phonon scattering rates as a function of period length and temperature to establish a more rigorous understanding of these effects.

3.3. Detailed Analysis of Superlattice Phonon Lifetimes

The linewidth of a SED peak corresponds directly to the phonon scattering rate, which is the inverse of the phonon lifetime. In this section, we examine how superlattice phonon lifetimes vary with both temperature and period length.

Figure 5a–c show the SED spectra for a transverse acoustic (TA) phonon mode at $k = 0.7644\ \mathrm{nm}^{-1}$ along the Γ–X direction of the 2UC-2UC superlattice. As the temperature increases, the SED peak broadens significantly. By fitting the SED peaks to Lorentzian functions, following the procedure detailed in Section 2, we extract phonon lifetimes of 53 ps, 38 ps, and 18 ps at 10 K, 30 K, and 100 K, respectively. A similar trend is observed in Figure 5d–f for another TA mode at $k = 0.735\ \mathrm{nm}^{-1}$ in the same direction, with the phonon lifetimes being 51 ps, 40 ps, and 17 ps at 10 K, 30 K, and 100 K, respectively. This systematic decrease in phonon lifetimes with increasing temperature arises from enhanced anharmonic phonon–phonon scattering, a well-established mechanism in crystalline solids [39,41]. This anharmonic effect not only reduces the lattice thermal conductivity of superlattices but also contributes to the loss of phonon coherence, as demonstrated in previous experimental and simulation studies [5,6,17,22,33]. A summary of the phonon lifetimes extracted at different temperatures for the TA branch of the 2UC–2UC superlattice is provided in Figure 5m, clearly illustrating a general trend of decreasing phonon lifetimes at higher temperatures.

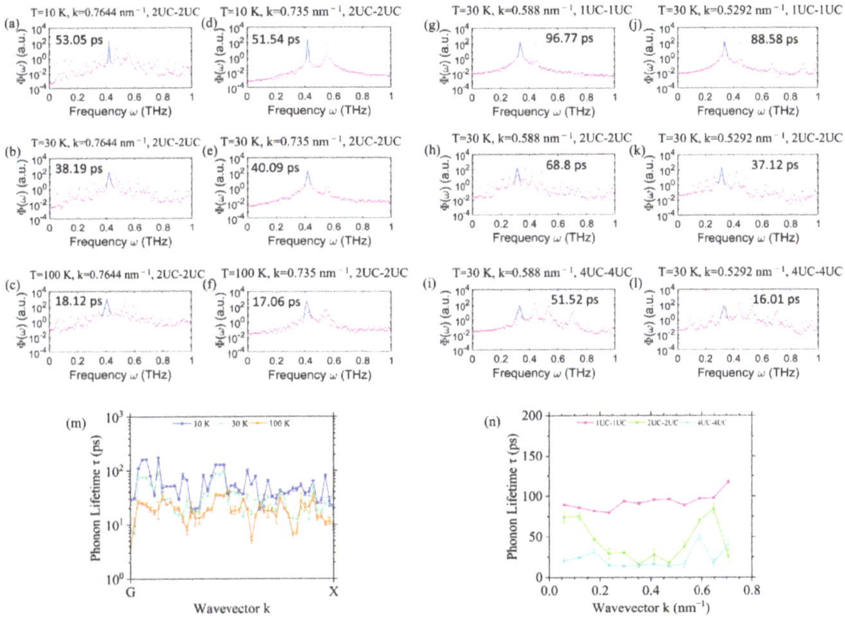

Figure 5. The effect of temperature and period length on the SED peak. (**a–c**) represent the effect of setting the temperature at 10 K, 30 K, and 100 K, respectively, on the 2UC-2UC structure at a wavevector $k = 0.7644\ \text{nm}^{-1}$. (**d–f**) show the effect of 10 K, 30 K, and 100 K temperatures, respectively, on the 2UC-2UC structure and at a wavevector $k = 0.735\ \text{nm}^{-1}$. (**g–i**) show the effect of the period length on the 1UC-1UC, 2UC-2UC, and 4UC-4UC structures, respectively, at 30 K and with a wavevector $k = 0.588\ \text{nm}^{-1}$, and (**j–l**) show the effect of the period length on the 1UC-1UC, 2UC-2UC, and 4UC-4UC structures, respectively, at 30 K and with a wavevector $k = 0.5292\ \text{nm}^{-1}$. (**m**) shows the lifetime of phonons in the 2UC-2UC superlattice structure at three different temperatures for the lowest acoustic phonon branch, and (**n**) shows the lifetime of phonons in the 1UC-1UC, 2UC-2UC, and 4UC-4UC superlattice structures at 30 K and with different wavevectors for the lowest acoustic phonon branch.

To investigate the influence of the period length on phonon lifetimes, we compare the same phonon modes across superlattices with different period lengths. Figure 5g–i present the SED peaks of a TA phonon mode at $k = 0.588\ \text{nm}^{-1}$ in 1UC-1UC, 2UC-2UC, and 4UC-4UC superlattices. The phonon lifetime decreases systematically with the increasing period length. Specifically, it is 97 ps, 69 ps, and 51 ps in 1UC-1UC, 2UC-2UC, and 4UC-4UC superlattices, respectively, showing a robust decreasing trend. The same trend is shown for another TA mode at $k = 0.5292\ \text{nm}^{-1}$ in Figure 5j–l. These findings are consistent with a recent anharmonic lattice dynamics study, which revealed shorter phonon lifetimes in superlattices with longer period lengths [29].

However, we emphasize that the reduction in phonon lifetime is not the sole contributor to the decreasing thermal conductivity observed in longer-period superlattices. As illustrated in Figure 4, the phonon dispersion relations undergo significant modifications due to the Brillouin zone folding (BZF) mechanism [10]. As a result, even phonon modes with identical reciprocal wavevectors in different superlattices exhibit differing frequencies. In superlattices with longer period lengths, these modes shift toward lower frequencies, reflecting the flattening of acoustic branches, which is an outcome predicted by early theoretical studies on BZF [6,9,10]. This flattening reduces the phonon group's velocity and, in combination with shorter lifetimes, leads to diminished thermal conductivity.

Figure 5n summarizes the extracted phonon lifetimes for the TA branch across the 1UC, 2UC, and 4UC superlattices. The trend confirms that increasing the superlattice period length leads to reduced phonon lifetimes. This behavior is expected, as longer period lengths enlarge the size of the superlattice unit cell and thus complicate the phonon dispersion landscape. This increased complexity enhances the available phase space for phonon–phonon scattering, which in turn elevates the scattering rate and reduces the phonons' lifetime.

3.4. *Decoherence of Coherent Superlattice Phonons at Higher Temperatures*

We emphasize that the phonon dispersion relations shown in this work correspond to the large unit cell of the superlattice, which represents one period of the superlattice. As is well established, these phonon dispersion relations can be understood as resulting from the folding of the original phonon dispersion curves of the base materials of the superlattice—namely, the m40 and m90 Lennard-Jones crystals, in this work. This phenomenon is known as the Brillouin zone folding mechanism [10]. Consequently, the phonon modes represented by the dispersion relations computed from the large unit cell are typically referred to as coherent phonons, since they are synchronized with the superlattice structure. Notably, these phonons do not perceive the interfaces within the superlattice as discontinuities, and, as a result, they are not scattered at the interfaces [16].

In contrast, incoherent phonons, which are associated with the respective base materials of the superlattice (m40 and m90), experience scattering at each interface. These incoherent phonons follow the dispersion relations of the pure m40 or m90 crystals. The distinction between coherent and incoherent phonons has important implications for the thermal transport properties of the superlattice [6,15,16,25]. Specifically, when thermal transport is dominated by coherent phonons, the thermal conductivity of the superlattice increases almost linearly with the number of periods (or equivalently, the total length of the structure). This behavior has been well documented in recent studies and is attributed to the ballistic transport of coherent phonons [6], which obey the dispersion relations of the superlattice rather than those of the individual base materials. On the other hand, when the thermal transport is dominated by incoherent phonons, the thermal conductivity of the superlattice remains nearly constant as more periods are added. This is because the incoherent, particle-like phonons are scattered at each interface, and the dense interfaces significantly limit the mean free path of the phonons. As a result, the thermal conductivity remains essentially constant with the increasing structure length.

In this work, we investigate a fundamental aspect of coherent superlattice phonons: how they maintain their coherence. Previous studies have indirectly observed the transition from coherent to incoherent phonon transport using nonequilibrium molecular dynamics simulations [15], as well as the coherent–incoherent two-phonon model proposed for superlattice structures [6]. These studies concluded that coherent phonons transition into incoherent modes at higher temperatures or when the superlattice period length increases. This conclusion was derived by extracting the overall thermal conductance from the thermal conductivity of periodic and aperiodic superlattices and fitting the results to the two-phonon model. In 2024 and 2025, we further used the atomistic phonon wavepacket method to directly demonstrate the conversion of incoherent phonons—those that follow the phonon dispersion relations of the base materials comprising the m40-m90 superlattice (similar to the superlattices studied in the present work)—into coherent phonon modes that follow the phonon dispersion relations of the superlattice [16,25]. Our results directly reveal that this conversion arises from coherent interference between phonon modes.

While these previous studies provided valuable insights into the mechanisms of coherent-to-incoherent phonon conversion, they were either inferred indirectly through molecular dynamics simulations or conducted at a background temperature of zero, where anharmonic phonon–phonon scattering is absent. In this work, we employ SED analysis to directly examine the energy stored in the superlattice's coherent modes, enabling us to determine whether these modes disappear at elevated temperatures.

As shown in Figure 6a, we select a high-frequency optical branch from the superlattice phonon dispersion relations to examine the decoherence of the coherent mode. Although this phonon branch is not expected to contribute significantly to heat transport in the superlattice, it resides in a distinct frequency range that lacks the degeneracy of other phonon modes. This allows us to unambiguously quantify the energy stored in this branch and, correspondingly, the phonon population.

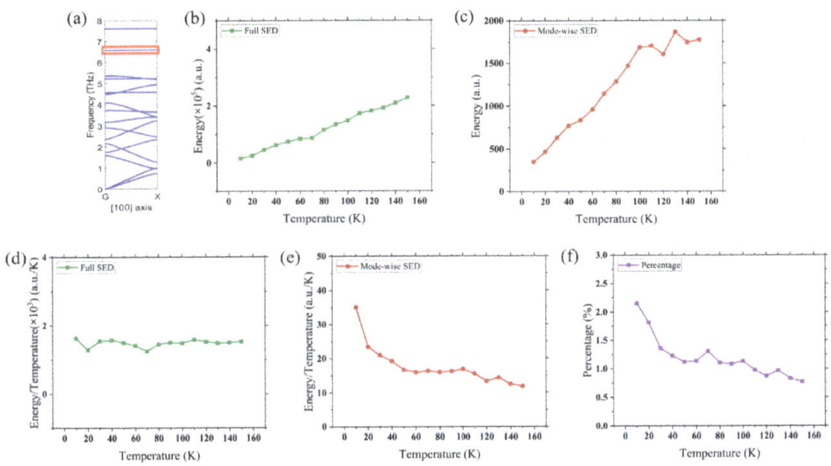

Figure 6. (**a**) Dispersion relation of the 2UC-2UC superlattice structure. (**b**) The full SED of the 2UC-2UC superlattice structure. (**c**) The mode-wise SED for the phonon branch indicated by the red rectangular box in panel (**a**), i.e., the selected optical branch. (**d**) The values of the full SED divided by temperature, (**e**) The mode-wise SED divided by temperature for the selected optical branch. (**f**) The mode-wise SED percentage of the selected optical branch with respect to the total SED value.

Figure 6b presents the total spectral energy density of all phonon modes across the entire phonon spectrum of the 2UC-2UC superlattice studied in this work. As expected, the total energy density increases monotonically with temperature. Figure 6c displays the total SED of the selected high-frequency optical branch in Figure 6a, which also exhibits an overall increasing trend with temperature. These results indicate that both the total thermal energy stored in the superlattice and that stored in the selected phonon branch increase with temperature.

In contrast, Figure 6d,e present the total spectral energy density normalized by temperature, which corresponds to the mode-wise heat capacity. This quantity serves as an indicator of the number of phonon modes available at a given temperature, which we will refer to as the total spectral heat capacity in subsequent discussions.

As shown in Figure 6d, the total spectral heat capacity of the entire phonon spectrum remains nearly constant with increasing temperature, with minor fluctuations arising from noise in the molecular dynamics simulations. This is expected, as the heat capacity remains at its classical limit ($3k_B$ per atom, where k_B is the Boltzmann constant) in classical molecular dynamics simulations.

Interestingly, Figure 6e shows that the total spectral heat capacity of the selected optical phonon branch in Figure 6a exhibits a pronounced decreasing trend with temperature. To further illustrate this behavior, we plotted the ratio of the total SED of the selected phonon branch to the total SED of the entire phonon spectrum in Figure 6f. This ratio also shows a significant decrease, indicating a loss of phonon modes in the selected phonon branch. This observation is critical, as it suggests that the optical phonon branch is losing phonon modes that store thermal energy. Since this branch belongs to the superlattice phonon dispersion relation, we infer that these superlattice phonon modes, or coherent modes, are converting into non-superlattice modes—specifically, incoherent modes that follow the phonon dispersion relations of the base materials. The reason for this is that at higher temperatures, the intensified anharmonic phonon scatterings, which break the phase and thus the coherence of phonons, hinder the formation of coherent phonon modes; instead, the vibration energy remains in the incoherent modes, which tend to follow the phonon dispersion relations of the local material, m40 or m90 in the m40-m90 superlattice studied in this work. This finding agrees with our recent study based on the two-phonon model, which suggests the conversion of coherent phonons into incoherent phonons at higher temperatures [15]. It also complements our recent atomistic phonon wavepacket study, which directly revealed the conversion of incoherent phonons into coherent phonons [16].

4. Conclusions

In this work, we conducted extensive molecular dynamics simulations to investigate the behavior of coherent superlattice phonon modes in multi-period superlattices composed of conceptual Lennard-Jones materials. By employing a phonon SED analysis, we elucidated the evolution of phonon modes associated with the superlattice phonon dispersion relation—commonly referred to as coherent phonons—as being a function of the temperature and superlattice period length. The SED method, which extracts detailed phonon properties from atomic trajectories, provides a realistic assessment of phonon dispersions and lifetimes. Our findings reveal that elevated temperatures significantly reduce the lifetimes of most superlattice phonon modes due to enhanced anharmonic phonon–phonon scattering. Notably, we observed a pronounced decrease in the spectral heat capacity of a high-frequency optical phonon branch at higher temperatures. This unambiguously indicates the decoherence of coherent phonon modes (consistent with our prior work that indirectly suggested the conversion of coherent modes), which obey the superlattice dispersion relations, into incoherent modes, which follow the dispersion relations of the constituent base materials. The intensified anharmonicity disrupts the phase coherence of the phonons, effectively destroying their coherent nature. While the complex structure of the superlattice phonon dispersion relation limits an unambiguous spectral heat capacity analysis of all phonon branches, our results offer strong support for the coherent-to-incoherent mode conversion mechanism existing under thermal excitation.

Furthermore, we examined the impact of the superlattice period length on phonon dispersions and lifetimes. In agreement with the Brillouin zone folding theory, the SED analysis revealed a significant flattening of acoustic branches with increasing period length, implying reduced phonon group velocities. This directly correlates with the observed reduction in lattice thermal conductivity in longer-period superlattices when coherent transport dominates. Interestingly, we also found that the lifetimes of many coherent phonon modes decrease as the superlattice period length increases, likely due to an expanded phase space for phonon–phonon scattering, further contributing to the suppression of thermal transport.

This study provides a comprehensive analysis of the temperature- and geometry-dependent behavior of coherent phonons in superlattices. Unlike previous investigations, which inferred the loss of phonon coherence indirectly through changes in thermal conductivity, our work presents direct evidence of the disappearance of coherent phonon modes, i.e., decoherence, at elevated temperatures. Using SED analysis, we demonstrate that this phenomenon arises from enhanced anharmonic scattering and coherence degradation. These results deepen our fundamental understanding of phonon transport in nanoscale periodic structures and offer valuable insights for the development of thermally tunable materials and devices.

Author Contributions: Conceptualization, M.N. and Y.W.; methodology, M.N. and Y.W.; software, M.N.; validation, M.N. and Y.W.; formal analysis, M.N. and Y.W.; investigation, M.N. and Y.W.; resources, Y.W.; data curation, M.N. and Y.W.; writing—original draft preparation, M.N. and Y.W.; writing—review and editing, M.N. and Y.W.; visualization, M.N.; supervision, Y.W.; project administration, Y.W.; funding acquisition, Y.W. All authors have read and agreed to the published version of the manuscript.

Funding: The authors gratefully acknowledge financial support from the National Science Foundation Thermal Transport Processes program under grant CBET-2047109. Nasiri's contributions to this project were also supported by the National Science Foundation Thermal Transport Processes program through grant CBET-1953300.

Data Availability Statement: The original contributions presented in this study are included in the article. Further inquiries can be directed to the corresponding authors.

Acknowledgments: The authors extend their gratitude to the Research and Innovation team and the Cyberinfrastructure Team in the Office of Information Technology at the University of Nevada, Reno, for providing access to the Pronghorn High-Performance Computing Cluster.

Conflicts of Interest: The authors declare no conflicts of interest. The funders had no role in the design of the study; in the collection, analyses, or interpretation of data; in the writing of the manuscript; or in the decision to publish the results.

References

1. Venkatasubramanian, R.; Siivola, E.; Colpitts, T.; O'quinn, B. Thin-film thermoelectric devices with high room-temperature figures of merit. *Nature* **2001**, *413*, 597–602. [CrossRef] [PubMed]
2. Harman, T.; Taylor, P.; Walsh, M.; LaForge, B. Quantum dot superlattice thermoelectric materials and devices. *Science* **2002**, *297*, 2229–2232. [CrossRef]
3. Chowdhury, I.; Prasher, R.; Lofgreen, K.; Chrysler, G.; Narasimhan, S.; Mahajan, R.; Koester, D.; Alley, R.; Venkatasubramanian, R. On-chip cooling by superlattice-based thin-film thermoelectrics. *Nat. Nanotechnol.* **2009**, *4*, 235–238. [CrossRef] [PubMed]
4. Bulman, G.; Barletta, P.; Lewis, J.; Baldasaro, N.; Manno, M.; Bar-Cohen, A.; Yang, B. Superlattice-based thin-film thermoelectric modules with high cooling fluxes. *Nat. Commun.* **2016**, *7*, 10302. [CrossRef] [PubMed]
5. Luckyanova, M.N.; Garg, J.; Esfarjani, K.; Jandl, A.; Bulsara, M.T.; Schmidt, A.J.; Minnich, A.J.; Chen, S.; Dresselhaus, M.S.; Ren, Z.; et al. Coherent phonon heat conduction in superlattices. *Science* **2012**, *338*, 936–939. [CrossRef]
6. Wang, Y.; Huang, H.; Ruan, X. Decomposition of coherent and incoherent phonon conduction in superlattices and random multilayers. *Phys. Rev. B* **2014**, *90*, 165406. [CrossRef]
7. Xie, G.; Ding, D.; Zhang, G. Phonon coherence and its effect on thermal conductivity of nanostructures. *Adv. Phys. X* **2018**, *3*, 1480417. [CrossRef]
8. Anufriev, R.; Maire, J.; Nomura, M. Review of coherent phonon and heat transport control in one-dimensional phononic crystals at nanoscale. *APL Mater.* **2021**, *9*, 070701. [CrossRef]
9. Nomura, M.; Anufriev, R.; Zhang, Z.; Maire, J.; Guo, Y.; Yanagisawa, R.; Volz, S. Review of thermal transport in phononic crystals. *Mater. Today Phys.* **2022**, *22*, 100613. [CrossRef]
10. Simkin, M.; Mahan, G. Minimum thermal conductivity of superlattices. *Phys. Rev. Lett.* **2000**, *84*, 927. [CrossRef]

11. Chen, Y.; Li, D.; Lukes, J.R.; Ni, Z.; Chen, M. Minimum superlattice thermal conductivity from molecular dynamics. *Phys. Rev. B—Condens. Matter Mater. Phys.* **2005**, *72*, 174302. [CrossRef]
12. Ravichandran, J.; Yadav, A.K.; Cheaito, R.; Rossen, P.B.; Soukiassian, A.; Suresha, S.; Duda, J.C.; Foley, B.M.; Lee, C.H.; Zhu, Y.; et al. Crossover from incoherent to coherent phonon scattering in epitaxial oxide superlattices. *Nat. Mater.* **2014**, *13*, 168–172. [CrossRef]
13. Wang, Y.; Gu, C.; Ruan, X. Optimization of the random multilayer structure to break the random-alloy limit of thermal conductivity. *Appl. Phys. Lett.* **2015**, *106*, 073104. [CrossRef]
14. Ma, T.; Lin, C.T.; Wang, Y. The dimensionality effect on phonon localization in graphene/hexagonal boron nitride superlattices. *2D Mater.* **2020**, *7*, 035029. [CrossRef]
15. Chakraborty, P.; Chiu, I.A.; Ma, T.; Wang, Y. Complex temperature dependence of coherent and incoherent lattice thermal transport in superlattices. *Nanotechnology* **2020**, *32*, 065401. [CrossRef] [PubMed]
16. Maranets, T.; Wang, Y. Prominent phonon transmission across aperiodic superlattice through coherent mode-conversion. *Appl. Phys. Lett.* **2024**, *125*, 042205. [CrossRef]
17. Cui, H.; Maranets, T.; Ma, T.; Wang, Y. Spectral heat flux redistribution upon interfacial transmission. *J. Phys. Condens. Matter* **2025**, *37*, 115002. [CrossRef]
18. Cingolani, R.; Lomascolo, M.; Lovergine, N.; Dabbicco, M.; Ferrara, M.; Suemune, I. Excitonic properties of ZnSe/ZnSeS superlattices. *Appl. Phys. Lett.* **1994**, *64*, 2439–2441. [CrossRef]
19. Prete, P.; Wolf, D.; Marzo, F.; Lovergine, N. Nanoscale spectroscopic imaging of GaAs-AlGaAs quantum well tube nanowires: Correlating luminescence with nanowire size and inner multishell structure. *Nanophotonics* **2019**, *8*, 1567–1577. [CrossRef]
20. Luckyanova, M.N.; Mendoza, J.; Lu, H.; Song, B.; Huang, S.; Zhou, J.; Li, M.; Dong, Y.; Zhou, H.; Garlow, J.; et al. Phonon localization in heat conduction. *Sci. Adv.* **2018**, *4*, eaat9460. [CrossRef]
21. Maranets, T.; Wang, Y. How phonon coherence develops and contributes to heat conduction in periodic and aperiodic superlattices. *arXiv* **2024**, arXiv:2412.15354. [CrossRef]
22. Chakraborty, P.; Cao, L.; Wang, Y. Ultralow lattice thermal conductivity of the random multilayer structure with lattice imperfections. *Sci. Rep.* **2017**, *7*, 8134. [CrossRef]
23. Maranets, T.; Doe, E.; Wang, Y. Role of interface mixing on coherent heat conduction in periodic and aperiodic superlattices. *arXiv* **2025**, arXiv:2501.17346. [CrossRef]
24. Chakraborty, P.; Liu, Y.; Ma, T.; Guo, X.; Cao, L.; Hu, R.; Wang, Y. Quenching thermal transport in aperiodic superlattices: A molecular dynamics and machine learning study. *ACS Appl. Mater. Interfaces* **2020**, *12*, 8795–8804. [CrossRef] [PubMed]
25. Maranets, T.; Nasiri, M.; Wang, Y. Influence of spatial coherence on phonon transmission across aperiodically arranged interfaces. *Phys. Lett. A* **2024**, *512*, 129572. [CrossRef]
26. Hu, R.; Iwamoto, S.; Feng, L.; Ju, S.; Hu, S.; Ohnishi, M.; Nagai, N.; Hirakawa, K.; Shiomi, J. Machine-learning-optimized aperiodic superlattice minimizes coherent phonon heat conduction. *Phys. Rev. X* **2020**, *10*, 021050. [CrossRef]
27. Ferrando-Villalba, P.; Chen, S.; Lopeandía, A.; Álvarez, F.X.; Alonso, M.; Garriga, M.; Santiso, J.; García, G.; Goñi, A.R.; Donadio, D.; et al. Beating the Thermal Conductivity Alloy Limit Using Long-Period Compositionally Graded Si1−x Ge x Superlattices. *J. Phys. Chem. C* **2020**, *124*, 19864–19872. [CrossRef]
28. Landry, E.; McGaughey, A. Effect of interfacial species mixing on phonon transport in semiconductor superlattices. *Phys. Rev. B—Condens. Matter Mater. Phys.* **2009**, *79*, 075316. [CrossRef]
29. Phonon transport governed by intrinsic scattering in short-period AlN/GaN superlattices. *Phys. Rev. B* **2024**, *109*, 104310. [CrossRef]
30. Thompson, A.P.; Aktulga, H.M.; Berger, R.; Bolintineanu, D.S.; Brown, W.M.; Crozier, P.S.; In't Veld, P.J.; Kohlmeyer, A.; Moore, S.G.; Nguyen, T.D.; et al. LAMMPS-a flexible simulation tool for particle-based materials modeling at the atomic, meso, and continuum scales. *Comput. Phys. Commun.* **2022**, *271*, 108171. [CrossRef]
31. Lee, S.M.; Cahill, D.G.; Venkatasubramanian, R. Thermal conductivity of Si–Ge superlattices. *Appl. Phys. Lett.* **1997**, *70*, 2957–2959. [CrossRef]
32. Chakraborty, S.; Kleint, C.; Heinrich, A.; Schneider, C.; Schumann, J.; Falke, M.; Teichert, S. Thermal conductivity in strain symmetrized Si/Ge superlattices on Si (111). *Appl. Phys. Lett.* **2003**, *83*, 4184–4186. [CrossRef]
33. Cui, H.; Maranets, T.; Ma, T.; Wang, Y. Spectral analysis of coherent and incoherent phonon transport in silicon nanomeshes. *Phys. Rev. B* **2024**, *110*, 075301. [CrossRef]
34. Tadano, T.; Gohda, Y.; Tsuneyuki, S. Anharmonic force constants extracted from first-principles molecular dynamics: applications to heat transfer simulations. *J. Phys. Condens. Matter* **2014**, *26*, 225402. [CrossRef]
35. Thomas, J.A.; Turney, J.E.; Iutzi, R.M.; Amon, C.H.; McGaughey, A.J. Predicting phonon dispersion relations and lifetimes from the spectral energy density. *Phys. Rev. B—Condens. Matter Mater. Phys.* **2010**, *81*, 081411. [CrossRef]
36. Panneerselvam, I.R.; Cui, H.; Maranets, T.; Wang, Y. Disorder-dominated and scattering-dominated thermal transport in clathrate hydrates. *Comput. Mater. Sci.* **2024**, *244*, 113189. [CrossRef]

37. Zhang, Z.; Guo, Y.; Bescond, M.; Chen, J.; Nomura, M.; Volz, S. Coherent thermal transport in nano-phononic crystals: An overview. *APL Mater.* **2021**, *9*, 081102. [CrossRef]
38. Zhang, Z.; Guo, Y.; Bescond, M.; Chen, J.; Nomura, M.; Volz, S. Heat conduction theory including phonon coherence. *Phys. Rev. Lett.* **2022**, *128*, 015901. [CrossRef]
39. Ma, T.; Chakraborty, P.; Guo, X.; Cao, L.; Wang, Y. First-principles modeling of thermal transport in materials: Achievements, opportunities, and challenges. *Int. J. Thermophys.* **2020**, *41*, 9. [CrossRef]
40. Chakraborty, P.; Xiong, G.; Cao, L.; Wang, Y. Lattice thermal transport in superhard hexagonal diamond and wurtzite boron nitride: A comparative study with cubic diamond and cubic boron nitride. *Carbon* **2018**, *139*, 85–93. [CrossRef]
41. Callaway, J. Model for lattice thermal conductivity at low temperatures. *Phys. Rev.* **1959**, *113*, 1046. [CrossRef]

crystals

MDPI

Article

A Compact and Fast Resonant Cavity-Based Encoder in Photonic Crystal Platform

Mohammad Soroosh [1,*], Faris K. AL-Shammri [2], Mohammad Javad Maleki [1,*], Venkatachalam Rajarajan Balaji [3] and Ehsan Adibnia [4]

1 Department of Electrical Engineering, Shahid Chamran University of Ahvaz, Ahvaz 6135783151, Iran
2 Biomedical Engineering Department, College of Engineering, University of Warith Al Anbiyaa, Karbala 56001, Iraq; faris.kar@uowa.edu.iq
3 Centre for Healthcare Advancement, Innovation and Research, Vellore Institute of Technology, Chennai 600127, Tamil Nadu, India; balaji.vr@vit.ac.in
4 Faculty of Electrical and Computer Engineering, University of Sistan and Baluchestan (USB), Zahedan P.O. Box 9816745563, Iran; ehsan.adibnia@pgs.usb.ac.ir
* Correspondence: m.soroosh@scu.ac.ir (M.S.); mj.maleki@scu.ac.ir (M.J.M.); Tel.: +98-061-33226602 (M.S. & M.J.M.)

Abstract: A novel 4-to-2 photonic crystal encoder is proposed by modulating the intensity of four input optical signals, and four distinct output states are achieved. Nonlinear rods are employed to couple input waves into resonant cavities, directing the light to the desired output waveguides. The proposed design, with a footprint of 114 μm^2, demonstrates efficient encoding operation at a wavelength of 1550 nm and is highly suitable for integrated photonics applications. A comprehensive comparative analysis revealed that the proposed 4-to-2 encoder exhibits a time response 176 fs faster than previously presented encoders. Furthermore, the contrast ratio of the designed structure is as high as 13.78 dB to distinguish between logic 0 and 1. These advancements hold significant potential for enhancing the performance of compact, high-speed digital circuits.

Keywords: contrast ratio; encoder; optical Kerr effect; photonic crystal; resonant cavity

1. Introduction

Optical processing offers a promising pathway to faster, more efficient computing. By using the unique properties of light, these systems replace electrons with photons, leading to significant improvements in speed and performance. One of the most notable advantages of optical processing is the incredibly high speed at which data can be transmitted. Photons travel near the speed of light, enabling optical systems to transfer vast amounts of data in a fraction of the time required by electronic counterparts [1]. Moreover, the immense bandwidth of optical technology allows for the simultaneous processing of multiple complex datasets. Other benefits include lower power consumption, reduced electromagnetic interference, and greater noise immunity. The non-material nature of light enables the creation of tiny and dense circuits, resulting in smaller, more efficient electronic devices. Optical processing can revolutionize fields such as quantum computing, artificial intelligence, and optical communications [2].

Photonic crystals are engineered structures that exhibit periodic variation in the refractive index. This periodic arrangement creates a lattice-like structure that influences the behavior of photons. In simpler terms, photonic crystals act like conventional crystals for electrons, but instead of controlling electrons, they manipulate the flow of photons [3]. These structures possess bandgaps, energy ranges where light propagation with specific frequencies is forbidden. This property gives rise to intriguing and beneficial optical characteristics. One of the most significant advantages of photonic crystals is their ability to control light precisely. By carefully designing the crystal structure, light can be directed along specific paths, confined to particular regions, or completely blocked. These capabilities are invaluable in the design of optical devices. Moreover, photonic crystals can serve as exact optical filters, permitting the transmission of only specific wavelengths. This property has extensive applications in spectroscopy and optical sensing. Additionally, photonic crystals can function as efficient waveguides and optical cavities, essential components in optical devices.

Adopting photonic crystal cavities instead of photonic crystal rings has led to significant advancements in optical device design. By enabling tighter confinement of light, higher quality factors, and greater design flexibility, these cavities have revolutionized the design of optical devices. Additionally, photonic crystal cavities offer reduced losses and the ability to create more intricate structures, opening up new possibilities for advanced optical functionalities. Photonic crystals offer unparalleled control over light, thanks to their precise periodic structures. This capability has paved the way for the design and fabrication of a wide range of optical devices with small sizes and high speeds. Encoders [4–14], decoders [15,16], adders [17,18], flip-flops [19–21], demultiplexers/multiplexers [22,23], and advanced optical sensors [24–26] are all examples of the potential applications of photonic crystals.

An encoder is a digital circuit that converts input binary data into a coded output, encoding 2n input lines into n bits. By exploiting the periodic structure of photonic crystals, all-optical encoders can perform encoding operations directly on optical signals. This eliminates the bottleneck associated with optoelectronic conversion, enabling high-speed and low-power optical information processing. Makvandi et al. [4] presented an all-optical 4-to-2 encoder based on a silicon photonic crystal. This device employed defect engineering to create cross-connections between four input ports and two output ports. By carefully arranging rods of different radii in the cross-connection area, the encoder effectively coupled optical signals. It features a compact footprint of only 133 μm^2 and offers a rapid switching speed with a maximum rise time of 205 fs. Additionally, the device demonstrated performance margins of 2% for logic 0 and 34% for logic 1. However, the incorporation of various defects complicates fabrication due to the need for multiple masks. Moreover, the encoder's performance is highly sensitive to the positioning of the rods, which introduces further challenges related to fabrication errors.

A photonic crystal encoder intended for optical computing was developed by Rajasekar et al. [5]. Their design, which is based on a barium titanate nanostructure platform, employs a dual nanocavity-coupled ring resonator, reflector, and waveguides. The operation of the device is grounded in interference and resonance principles, allowing for a response time of 369.3 fs, a compact area of 174.24 μm^2, and a contrast ratio of 7.11 dB. However, for the encoding process, they assumed different phases for the operational states, which may not be feasible in a practical optical circuit. A silica-based photonic crystal fiber (PCF) was introduced by Naghizade et al. for an all-optical 4-to-2 encoder [6]. The device's architecture, which includes an OR gate and a buffer, is designed for operation in the C-band. Numerical simulations indicated that the encoder achieved normalized transmission values of less than 0.02% for logic 0 and more than 95% for logic 1. It also demonstrated a delay time of 0.3 ps and occupied an area of 1500 μm^2. However, the large area of the designed structure is a significant drawback, making it unsuitable for optical integrated circuits. Fallahi et al. [7] introduced a 4-to-2 encoder that employs a wave interference technique based on 2D photonic crystal structures. To reduce losses and scattering, the design incorporates nano-resonators and curved two-branch waveguides, resulting in a compact footprint of 149 μm^2. Photonic bandgap and output spectrum analyses were conducted using plane wave expansion (PWE) and finite difference time domain (FDTD) simulations, respectively. The device achieved a contrast ratio of 7.88 dB, a delay time of 0.21 ps, and a bit rate of 4.761 Tb/s at a wavelength of 1550 nm. However, similar to reference [5], the reliance on different phases for operational states restricts the potential applications of the designed structure. An all-optical 4-to-2 encoder based on a ring resonator was realized by Veisi et al. [8]. To analyze the eigenvalues, they employed the PWE method. The final design features a footprint of 182 μm^2, achieving a contrast ratio of 27.78 dB and a delay time of 250 fs at a central wavelength of 1555 nm, which allows for a bit rate of 4 Tb/s. While they were able to enhance the contrast ratio through strong coupling in the resonant ring, the ring-based design resulted in a larger overall area.

Kamal et al. [9] introduced a two-dimensional 4-to-2 optical encoder based on photonic crystal structures. This device, which features silicon rods in an air background, utilizes wave interference for its operation. To analyze the photonic bandgap and electric field distribution, they employed the PWE and FDTD methods, respectively. The encoder has a footprint of 204.8 μm^2, a contrast ratio of 6.69 dB, and a response time of 254 fs. It operates at a wavelength of 1550 nm, making it compatible with optical communication systems. However, the low contrast ratio may limit the device's suitability for applications in optical circuits, as adjacent coupled circuits might struggle to differentiate between logic 0 and 1. A digital all-optical 4×2 encoder architecture based on a hexagonal lattice of silicon rods surrounded by air cladding was introduced by Askarian [10]. This device operates on the principle of threshold switching, utilizing the optical Kerr effect in combination with nonlinear ring resonators. FDTD simulations indicated a performance with a contrast ratio of 16.98 dB, a response time of 1.5 ps, and a switching speed of 667 Gbit/s at a wavelength of 1550 nm. However, the use of the Kerr effect requires high input power, which is unsuitable for integrated circuits and can lead to undesirable phenomena in the nonlinear regime. Additionally, Arunkumar et al. [11] designed a silicon photonic crystal 4×2 encoder based on a divider and nanocavities created by defects. This structure, optimized for a wavelength of 1550 nm, features a compact footprint of 295 μm^2, a delay time of over 170 fs, a high contrast ratio of 31.14 dB, a fast response time of 240 fs, low cross-talk of −31.14 dB, and low loss of −3.18 dB. Similar to [5,7], the design requires different phases to achieve correct operation under varying input states, which may limit the encoder's applicability in optical circuits.

Bouaouina and Lebbal [12] reported a structure based on nonlinear resonant rings that performed 4-input to 2-output encoding using the optical Kerr effect. Although they reported a contrast ratio of 37.02 dB and a bit rate of 2.23 Tb/s, the high power required to achieve the Kerr effect is a major problem with this structure. On the other hand, the use of polymer rods with different radii severely reduces the fabrication feasibility. Therefore, the idea of using it in a practical situation does not seem to be feasible. The use of a one-dimensional photonic crystal structure consisting of air holes in the vicinity of a waveguide can change the light transmission efficiency of the waveguide [13]. It was shown that if a stack of graphene and Al_2O_3 is used in the central hole of the photonic crystal, two ON and OFF states can be defined for the waveguide by adjusting the chemical potential of graphene between 0.2 eV and 0.8 eV. Accordingly, a 4-by-2 optical encoder was designed with an area of 127 μm^2 and a contrast ratio of 7.6 dB. This structure is electro-optical, and an electric voltage needs to be used to adjust the chemical potential of graphene. Switching between the ON and OFF states severely limits the operating frequency of the structure. By using the interference through a resonant ring and multiple defects, it was shown that 4-to-2 encoding can be realized [14]. The use of closely spaced waveguides led Mostafa et al. to propose a photonic crystal-based structure with an area of 128 μm^2 and report a delay time of the structure to the input signal excitation of 0.1 ps. Although the designed structure showed a good time response, its contrast ratio was 7.11 dB, which is not suitable for coupling to subsequent circuits. The close proximity of the waveguides increases the crosstalk of the structure.

According to previous studies, it is evident that the performance of photonic crystal-based encoders can be improved by minimizing their physical sizes and enhancing their response speed. With this in mind, this research introduces an innovative all-optical 4-to-2 encoder that uses a photonic crystal platform. The proposed design offers a significantly smaller footprint than its predecessors, making it a promising candidate for optical device integration. Furthermore, given the growing need for high-speed data processing, another primary goal of this research is to reduce the device's response time. Simulation results demonstrate that the proposed structure surpasses previous designs regarding response time. The structural parameters of the proposed device have been designed to achieve superior performance. The second part of this study presents the proposed structure and its simulation, providing a detailed account of the simulation methodologies employed. Finally, a concise summary of the key findings is provided.

2. The Proposed Structure

The proposed structure includes four inputs (I_0, I_1, I_2, and I_3), two outputs (O_0 and O_1), and eight waveguides (W1 to W8), as illustrated in Figure 1. The encoder comprises an array of dielectric rods with n = 3.1, formed in a square arrangement within an air cladding. The array consists of 22 rods along the x-axis and 20 rods along the z-axis, resulting in a footprint of 114 μm^2. The rod radius, r, is set to 0.23a, where a represents the lattice constant of 510 nm. Eight defects are introduced at the waveguide intersections to enhance light coupling into the waveguides. The red disks represent the fundamental rods, and the small red disks refer to the defects. The blue and yellow disks show the nonlinear rods. Defects in photonic crystal structures are crucial in guiding and manipulating light waves. Point defects create localized resonant cavities that trap light, enhancing light–matter interactions and enabling filters. Additional structural parameters are summarized in Table 1.

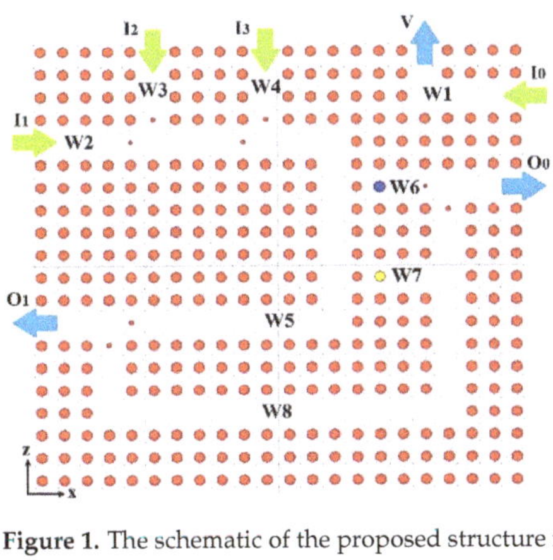

Figure 1. The schematic of the proposed structure in the xz-plane including input ports I_0 to I_3 and output ports O_0 and O_1.

Table 1. The structural parameters of the proposed encoder.

Parameter	Symbol	Value	Unit
The lattice constant	a	510	nm
The radius of fundamental rods	r	117	nm
The refractive index of fundamental rods	n	3.1	-
The radius of blue rods	r_b	124	nm
The refractive index of blue rods	n_b	1.4	-
The radius of yellow rods	r_y	114	nm
The refractive index of yellow rods	n_y	1.4	-

The radius of the rods plays a crucial role in the photonic crystal's ability to generate a photonic bandgap. Specifically, a larger rod radius can enhance light interaction, potentially resulting in a broader bandgap. Additionally, the size and shape of the rods impact how light is localized within the crystal, affecting both the quality factor of the modes and the light transmission characteristics. The lattice constant, which dictates the periodicity of the photonic crystal, is essential for bandgap formation. Careful selection of the rod arrangement and spacing is necessary to achieve the desired symmetry and efficiency in light manipulation. A larger lattice constant allows longer wavelengths to interact with the crystal, influencing the propagation of different wavelengths, while a smaller lattice constant is more effective for shorter wavelengths. Moreover, the lattice constant is critical for integrating photonic crystals with other optical components, ensuring that the dimensions and operational wavelengths are compatible. Although decreasing the lattice constant and rod radius facilitates the reduction of device size, fabrication limitations must also be taken into account. According to fabrication reports, the fundamental radius and lattice constant were explored within the ranges of 45–160 nm and 380–630 nm, respectively, with proper results achieved at 117 nm for the rod radius and 510 nm for the lattice constant.

To couple light into the desired output ports, two nonlinear cavities with different radii (blue and yellow rods) are employed. The blue rod has a radius of 124 nm and the yellow rod has a radius of 114 nm. Both rods are made of doped glass with a refractive index of 1.4 and a nonlinear coefficient of 10–16 m^2/W [27]. Concerning Bragg's principle, forward waves with wavelength λ and backward waves from two periodic layers are in phase if Bragg's condition, $n_1d_2 + n_2d_2 = \lambda/2$, is fulfilled [28]. Rod and air gap are considered two periodic layers with refractive indices of n_1 and n_2 and thicknesses of d_1 and d_2. By changing the radius of the nonlinear rod, this condition is met, and consequently, the light transmission path is directed toward the desired outputs.

Four working cases for a 4-to-2 encoder are possible at a single input active regime. When only input I_0 is active ($I_3 = I_2 = I_1 = 0$, $I_0 = 1$), the output binary code of 00 is generated on the output lines O_1 and O_0. When only input I_1 equals 1 ($I_3 = I_2 = I_0 = 0$, $I_1 = 1$), the encoder outputs 01 on the output lines O_1 and O_0 are shown. For $I_2 = 1$ and $I_3 = I_1 = I_0 = 0$, outputs O_1 and O_0 provide code 10. If $I_3 = 1$ and the other inputs equal 0, the encoder provides code 11 on the output lines O_1 and O_0. In summary, the 4-to-2 encoder maps the active input line to its corresponding binary code on the output lines.

Using the considered values for the rod radii and lattice constant, the photonic band gap (PBG) is shown in Figure 2. In this study, the PWE method is used to calculate the photonic bandgap. Maxwell's equations are given by Equations (1) and (2) [29]:

$$\frac{1}{\varepsilon_r}\nabla \times \nabla \times E = \left(\frac{\omega}{c}\right)^2 E \quad (1)$$

$$\nabla \times \frac{1}{\varepsilon_r}\nabla \times H = \left(\frac{\omega}{c}\right)^2 H \quad (2)$$

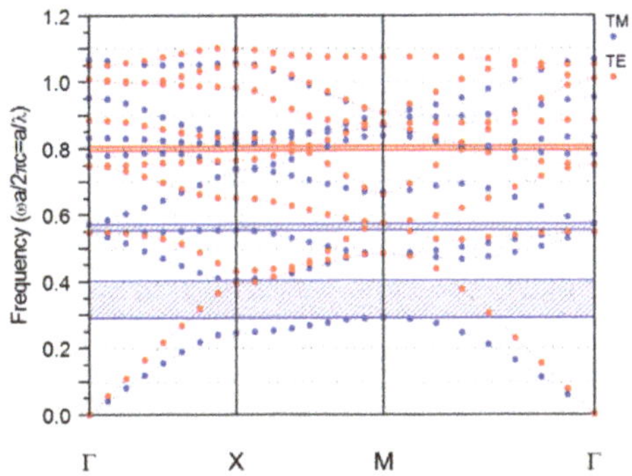

Figure 2. The band diagram of the designed structure. The structure provides two PBGs in TM mode and one PBG in TE mode. Dashed regions represent the photonic band gaps.

Here, ε_r represents the relative permittivity, c refers to the speed of light, and ω is the optical frequency. If the electric and magnetic fields are given by Fourier series expansions, $(\omega/c)^2$ satisfies the mentioned equations as the eigenvalue. According to Figure 2, the structure has two band gaps in the transverse magnetic (TM) mode and one band gap in the transverse electric (TE) mode. The band gaps in the TM mode correspond to the frequency ranges $0.289 < a/\lambda < 0.402$ and $0.552 < a/\lambda < 0.571$, where the first band gap is equivalent to the wavelength range $1269\ \text{nm} < \lambda < 1765\ \text{nm}$, which sufficiently covers the third telecommunication window. Therefore, the wavelength of 1550 nm is considered for simulation.

3. Results

To calculate the components of electric and magnetic fields, the FDTD method is employed. This numerical technique discretizes Equations (1) and (2) in time and space, enabling the calculation of field components. Perfect-matched layer (PML) boundary conditions are applied to absorb outgoing waves. The spatial discretization, with cell sizes Δx and Δz of 0.2 nm, is finer than 0.1λ. The Courant–Friedrichs–Lewy condition is a crucial criterion in numerical analysis, particularly in finite difference time domain simulations. It ensures the stability and accuracy of the numerical solution by limiting the time step size (Δt) relative to the spatial discretization [30].

$$c\Delta t \le \frac{1}{\sqrt{\left(\frac{1}{\Delta x^2} + \frac{1}{\Delta z^2}\right)}} \quad (3)$$

In this study, an intensity of $I = 100\ \text{mW}/\mu\text{m}^2$ is applied to ports I_0 and I_1, while 2I and 3I are entered in I_2 and I_3, respectively, to activate them. These values cause the difference in light intensity reaching the nonlinear rods. Thus, the coupling through cavities depends on the incoming intensity. Therefore, distinguishing among the working cases is possible, and the structure provides a binary code at outputs O_1 and O_0 in response to the activated input. Concerning the inputs of the structure and the single active regime, four working cases are possible, as follows:

Case #1: When input I_0 is active, light with intensity I enters the structure, activating port V. All other ports are in the logic 0 ($O_1 = O_0 = 0$) (as shown in Figure 3a).

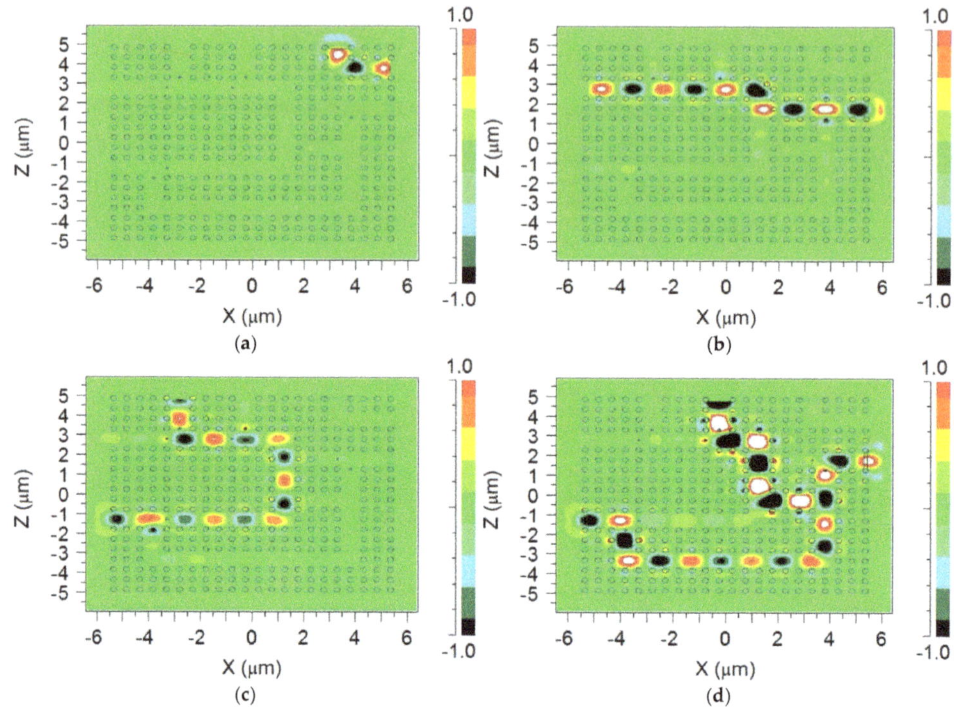

Figure 3. The electric field distribution for working case (**a**) $I_3I_2I_1I_0$ = 0001, (**b**) $I_3I_2I_1I_0$ = 0010, (**c**) $I_3I_2I_1I_0$ = 0100, and (**d**) $I_3I_2I_1I_0$ = 1000.

Case #2: If input I_1 is active and all other input ports are inactive, light is coupled into the structure through waveguide W2. Given that the input light intensity is I, the resonant condition is satisfied in the cavity including the blue rod, and light is transmitted through this cavity and waveguide W6 toward output O_0, so O_1 = 0 and O_0 = 1 (as illustrated in Figure 3b).

Case #3: When input I_2 is active, light with intensity 2I is coupled into the structure through waveguide W3. Due to the input light intensity, resonant conditions cannot be established in either cavity. Consequently, light is transmitted through waveguide W5 toward port O_1, becoming 1, and code 10 is generated at O_1O_0. (see Figure 3c).

Case #4: Finally, when light is coupled into the structure through input port I_3 with an intensity of 3I, resonant conditions are met in the cavity including the yellow rod. As a result, after coupling to waveguide W7, light is transmitted toward output ports O_1 and O_0, setting both output ports to logic 1 (as depicted in Figure 3d). As shown in Figure 3, the proposed structure successfully performs the encoding operation, and its correct operation can be verified.

A critical aspect of the proposed device is its temporal performance. In this work, the rise time as the time required for the output power to reach 90% of its steady-state value is calculated. Figure 4 depicts the temporal behavior of the encoder corresponding to a step trigger in inputs, and Table 2 summarizes its key characteristics. It can be seen that the rise time is equal to 106 fs, 122 fs, 138 fs, and 176 fs for $I_3I_2I_1I_0$ = 0001, $I_3I_2I_1I_0$ = 0010, $I_3I_2I_1I_0$ = 0100, and $I_3I_2I_1I_0$ = 1000, respectively. Moreover, the normalized power for logic 1 is 0.89, 0.77, 0.79, and 0.43 for cases 1, 2, 3, and 4, respectively. The normalized logic levels 0, 0.0001, 0.018, 0.014, and 0.001 correspond to 0001, 0010, 0100, and 1000. To calculate the normalized output power levels at O_1 and O_0 for mentioned cases, the worst-case scenarios are considered. Specifically, the minimum power level for logic 1 and the maximum power level for logic 0 are assumed to be the encoder's margins. As shown in Table 2, the normalized output power levels for logic 0 and 1 are 0.018 and 0.43, respectively. Additionally, the rise time of the proposed structure is 176 fs.

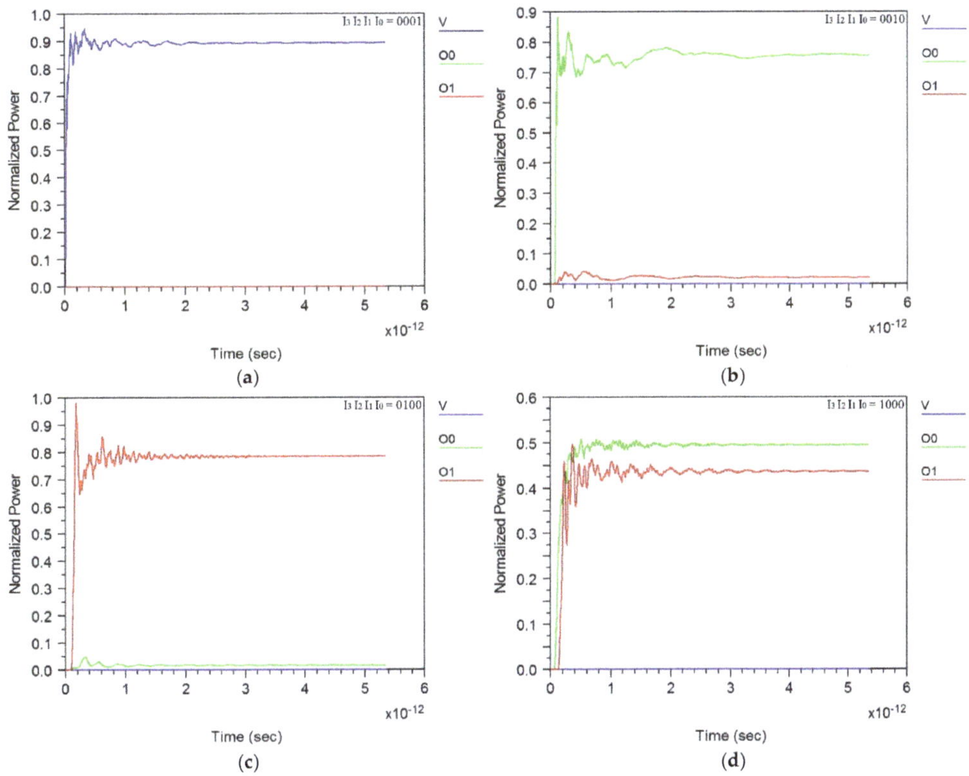

Figure 4. The normalized power at output ports in terms of time for different working cases (**a**) 0001, (**b**) 0010, (**c**) 0100, and (**d**) 1000.

Table 2. The time response of the designed structure.

Case	Input				Output			Margin	Transmitting Waveguides	Response Time (fs)
					Binary Code at O_1O_0	Normalized Power				
	I_3	I_2	I_1	I_0		O_1	O_0			
1	0	0	0	1	00	0	0	×	W1	106
2	0	0	1	0	01	0.018	0.77	M_0	W2, W6	122
3	0	1	0	0	10	0.79	0.014	×	W3, W5	138
4	1	0	0	0	11	0.43	0.49	M_1	W4, W7, W8	176

As detailed in Table 2, the calculated margins for logic states 0 and 1 were determined to be 0.018 and 0.43, respectively. This resulted in a margin difference of 0.412 and a contrast ratio (CR) of 13.78 dB, where the contrast ratio is calculated as [31]:

$$CR = 10\log(\frac{M_1}{M_0}) \quad (4)$$

To assess the performance of the proposed encoder, a comparative analysis of its critical features against those of previously reported encoders is given in Table 3. The designed structure occupies only 114 μm², which is smaller than all the structures previously presented. This feature is a significant advantage because the miniaturization of devices leads to an increase in the density of optical circuits. The response speed of the proposed structure is 176 fs, which is faster than that in other works (except [14]), according to Table 3. The increased response speed increases the data transfer rate, which is an attractive feature in high-speed processing. The contrast ratio of the designed encoder is 13.78 dB, which is higher than that in most previous works and shows that the proposed encoder can distinguish logic 1 and 0 well. By examining the area, time response, and contrast ratio of the proposed encoder in comparison with other structures, it can be concluded that this work improved the performance of all-optical encoders based on photonic crystals.

Table 3. The results obtained in this work, along with other photonic crystal works.

Reference	Year	Structure	Area (μm²)	Rise Time (fs)	CR (dB)
[4]	2020	Cross-connected waveguides	133	205	12.3
[5]	2020	Resonant ring	174.24	369	7.11
[6]	2020	Photonic crystal fiber	1500	300	13.81
[7]	2021	Resonant ring	149	210	7.88
[8]	2023	Resonant ring	182	250	27.78
[9]	2023	Resonant ring	204.8	254	6.69
[10]	2023	Resonant ring	410	1500	16.98
[11]	2024	Cross-connected waveguides	295	240	31.14
[12]	2023	Nonlinear Resonant ring	172	218	37.02
[13]	2020	Gaphene-Al_2O_3 Stack	127	1430	7.6
[14]	2019	Resonant ring	128	100	7.11
This work	2024	Resonant cavity	114	176	13.78

Many articles have reported on the fabrication of photonic crystal-based structures, such as [32–42]. They have used different methods to fabricate different devices, such as colloidal self-assembly, electron beam lithography, and direct writing via multiphoton microlithography. The deviation of the rods concerning the vertical axis, the ruggedness of the rods, and the non-periodicity of the lattice are the main challenges of the fabrication process. According to the reports, the fabrication of silicon rods with a radius of 45 nm and a spatial period of 380 nm is possible. The smallest radius of the rods in the presented structure is equal to 114 nm, and the lattice constant equals 510 nm.

4. Conclusions

In this paper, an all-optical 4-to-2 encoder was proposed. It was shown that different working cases can be realized if different powers are applied to the input ports. Using nonlinear rods makes it possible to generate binary codes corresponding to the inputs at the output ports. The small area of the designed structure (114 μm²) compared to other structures is one of the outstanding features required for optical processors. The response time of 176 fs shows the fast response of the presented device. Fast and compact devices are highly needed for optical circuits. Moreover, a contrast ratio of 13.78 dB helps to present an efficient encoding operation. The fast response, compact area, and high contrast ratio demonstrate an improvement in the performance of photonic crystal-based encoders.

Author Contributions: Conceptualization, M.J.M. and M.S.; methodology, F.K.A.-S. and M.J.M.; software, V.R.B. and M.J.M.; validation, E.A. and M.S.; formal analysis, M.J.M. and E.A.; investigation, M.S. and E.A.; writing—original draft preparation, M.J.M. and F.K.A.-S.; writing—review and editing, M.S., E.A. and V.R.B.; supervision, M.S.; project administration, M.S.; funding acquisition, M.S. All authors have read and agreed to the published version of the manuscript.

Funding: This work was supported by Shahid Chamran University of Ahvaz, grant number SCU.EE1403.672.

Data Availability Statement: All data generated or analyzed during this study are included in this published article.

Conflicts of Interest: The authors declare no conflicts of interest.

References

1. Soroosh, M.; Parandin, F.; Haddadan, F.; Maleki, M.J.; Bagheri, F. Highly efficient materials for photonic crystal-based optical components. In *Advances in All-Optical Communication*; Dhanabalan, S.S., Thirumurugan, A., Thirumaran, S., Eds.; IOP Publishing: London, UK, 2024; pp. 1–20.
2. Maleki, M.J.; Soroosh, M.; Parandin, F.; Haddadan, F. Photonic crystal-based decoders: Ideas and structures. In *Recent Advances and Trends in Photonic Crystal Technology*; IntechOpen: London, UK, 2023; pp. 91–108.
3. Maleki, M.J.; Soroosh, M.; Akbarizadeh, G.; Parandin, F.; Haddadan, F. Photonic Crystal Resonators in Designing Optical Decoders. *J. Optoelectron. Nanostruct.* **2023**, *8*, 1–24.
4. Makvandi, M.; Maleki, M.J.; Soroosh, M. Compact all-optical encoder based on silicon photonic crystal structure. *J. Appl. Res. Electr. Eng.* **2020**, *1*, 1–7.
5. Rajasekar, R.; Thavasi Raja, G.; Jayabarathan, J.K.; Robinson, S. High speed nano-optical encoder using photonic crystal ring resonator. *Photonic Netw. Commun.* **2020**, *40*, 31–39. [CrossRef]
6. Naghizade, S.; Saghaei, H. A novel design of all-optical 4 to 2 encoder with multiple defects in silica-based photonic crystal fiber. *Optik* **2020**, *222*, 165419. [CrossRef]
7. Fallahi, V.; Mohammadi, M.; Kordrostami, Z.; Seifouri, M.; Olyaee, S. Design and optimization of an ultra-fast symmetrical 4× 2 encoder based on 2D photonic crystal nano-resonators for integrated optical circuits. *Opt. Quantum Electron.* **2021**, *53*, 574. [CrossRef]
8. Veisi, E.; Mohammadi, M.; Seifouri, M.; Olyaee, S. Design and numerical analysis of high-performance all-optical 4 × 2 encoder using photonic crystal ring resonator. *Opt. Quantum Electron.* **2023**, *55*, 376. [CrossRef]
9. Kamal, S.M.; Ali, T.A.; Rafat, N.H. New designs of 4 × 2 photonic crystal encoders using ring resonators. *Opt. Quantum Electron.* **2023**, *55*, 261. [CrossRef]
10. Askarian, A. Design and implementation of all optical 4 × 2 encoder based on 2D-PhC platform and optical Kerr effect. *Opt. Quantum Electron.* **2023**, *55*, 822. [CrossRef]
11. Arunkumar, R.; Robinson, S. Design and Comparative Analysis of an Ultra-Fast, Low-Power All-Optical 4× 2 Encoder Using a Silicon Y-Shaped 2D Photonic Crystal. *Silicon* **2024**, *16*, 4997–5008. [CrossRef]
12. Bouaouina, M.S.; Lebbal, M.R. All optical photonic crystal encoder based on nonlinear Kerr effect with ultrahigh contrast ratio and low threshold power. *Opt. Quantum Electron.* **2023**, *55*, 867. [CrossRef]
13. Haddadan, F.; Soroosh, M.; Alaei-Sheini, N. Designing an electro-optical encoder based on photonic crystals using the graphene–Al_2O_3 stacks. *Appl. Opt.* **2020**, *59*, 2179–2185. [CrossRef] [PubMed]
14. Mostafa, T.S.; Mohammed, N.A.; El-Rabaie, E.S.M. Ultracompact ultrafast-switching-speed all-optical 4 × 2 encoder based on photonic crystal. *J. Comput. Electron.* **2019**, *18*, 279–292. [CrossRef]
15. Maleki, M.J.; Soroosh, M.; Mir, A. Improving the performance of 2-to-4 optical decoders based on photonic crystal structures. *Crystals* **2019**, *9*, 635. [CrossRef]
16. Askarian, A.; Parandin, F. Investigations of all optical 2 × 4 decoder based on PhC structure and nonlinear ring resonators. *Microw. Opt. Technol. Lett.* **2024**, *66*, e70004. [CrossRef]
17. Parandin, F. Ultra-compact and low delay time all optical half adder based on photonic crystals. *Opt. Quantum Electron.* **2023**, *55*, 398. [CrossRef]
18. Maleki, M.J.; Mir, A.; Soroosh, M. Ultra-fast all-optical full-adder based on nonlinear photonic crystal resonant cavities. *Photonic Netw. Commun.* **2021**, *41*, 93–101. [CrossRef]
19. Soma, S.; Sonth, M.V.; Gowre, S.C. Design of two-dimensional photonic crystal based ultra compact optical RS flip-flop. *Photonic Netw. Commun.* **2022**, *43*, 109–115. [CrossRef]

20. Parandin, F.; Sheykhian, A.; Askarian, A. A novel design of an all-optical D flip-flop based on 2D photonic crystals. *Microw. Opt. Technol. Lett.* **2024**, *66*, e34006. [CrossRef]
21. Zamanian-Dehkordi, S.S.; Soroosh, M.; Akbarizadeh, G. An ultra-fast all-optical RS flip-flop based on nonlinear photonic crystal structures. *Opt. Rev.* **2018**, *25*, 523–531. [CrossRef]
22. Maleki, M.J.; Soroosh, M. An ultra-fast all-optical 2-to-1 digital multiplexer based on photonic crystal ring resonators. *Opt. Quantum Electron.* **2022**, *54*, 397. [CrossRef]
23. Da Costa Tavares, S.C.; De Sousa, F.B.; De Oliveira, L.A.; De Sousa, F.M.; Miranda, I.R.S.; Costa, M.B. Four-channel photonic crystal demultiplexer with graphene with high quality factor for DWDM applications. *Opt. Quantum Electron.* **2024**, *56*, 622. [CrossRef]
24. Abd-ALhussain, M.J.; Rasheed, B.G.; Fakhri, M.A. Solid-core photonic crystal fiber-based nanolayer glucose sensor. *J. Opt.* **2024**, *53*, 2392–2404. [CrossRef]
25. Pravesh, R.; Kumar, D.; Pandey, B.P.; Chaudhary, V.S.; Singh, D.; Kumar, S. Advanced refractive index sensor based on photonic crystal fiber with elliptically split cores. *Opt. Quantum Electron.* **2023**, *55*, 1205. [CrossRef]
26. Gryga, M.; Ciprian, D.; Hlubina, P.; Pokorny, P.; Sobota, J. Narrow Tamm resonances in one-dimensional photonic crystals employed in sensor applications. *Opt. Laser Technol.* **2023**, *167*, 109797. [CrossRef]
27. Saleh, B.E.; Teich, M.C. *Fundamentals of Photonics*, 3rd ed.; John Wiley & Sons: Hoboken, NJ, USA, 2019.
28. Maleki, M.J.; Soroosh, M.; Mir, A. Ultra-fast all-optical 2-to-4 decoder based on a photonic crystal structure. *Appl. Opt.* **2020**, *59*, 5422–5428. [CrossRef]
29. Bao, G.; Li, P. *Maxwell's Equations in Periodic Structures*, 1st ed.; Springer: Singapore, 2022.
30. Hazra, S.; Mukhopadhyay, S. Photonic crystal based integrated system for half adder and half subtractor operations. *Opt. Quantum Electron.* **2024**, *56*, 855. [CrossRef]
31. Rafiee, E.; Afkhami, M. Design of an all-optical compact 2*1 multiplexer based on 2D photonic crystal ring resonators. *Opt. Quantum Electron.* **2024**, *56*, 283. [CrossRef]
32. Rechtsman, M.C. Reciprocal topological photonic crystals allow backscattering. *Nat. Photonics* **2023**, *17*, 383–384. [CrossRef]
33. Bommer, S.P.; Panuski, C.; Guilhabert, B.; Xia, Z.; Smith, J.A.; Dawson, M.D.; Englund, D.; Strain, M.J. Transfer printing micro-assembly of silicon photonic crystal cavity arrays: Beating the fabrication tolerance limit. *arXiv* **2024**, arXiv:2406.20010.
34. Li, J.; Yan, J.; Jiang, L.; Yu, J.; Guo, H.; Qu, L. Nanoscale multi-beam lithography of photonic crystals with ultrafast laser. *Light Sci. Appl.* **2023**, *12*, 164. [CrossRef]
35. Biswas, U.; Nayak, C.; Rakshit, J.K. Fabrication techniques and applications of two-dimensional photonic crystal: History and the present status. *Opt. Eng.* **2023**, *62*, 010901. [CrossRef]
36. Lowell, D.; Hassan, S.; Sale, O.; Adewole, M.; Hurley, N.; Philipose, U.; Chen, B.; Lin, Y. Holographic fabrication of graded photonic super-quasi-crystals with multiple-level gradients. *Appl. Opt.* **2018**, *57*, 6598–6604. [CrossRef] [PubMed]
37. Vyatskikh, A.; Ng, R.C.; Edwards, B.; Briggs, R.M.; Greer, J.R. Additive manufacturing of high-refractive-index, nanoarchitected titanium dioxide for 3D dielectric photonic crystals. *Nano Lett.* **2020**, *20*, 3513–3520. [CrossRef] [PubMed]
38. Abd-Elnaiem, A.M.; Mohamed, Z.E.A.; Elshahat, S.; Almokhtar, M.; Norek, M. Recent progress in the fabrication of photonic crystals based on porous anodic materials. *Energies* **2023**, *16*, 4032. [CrossRef]
39. Lowell, D.; Hassan, S.; Adewole, M.; Philipose, U.; Chen, B.; Lin, Y. Holographic fabrication of graded photonic super-crystals using an integrated spatial light modulator and reflective optical element laser projection system. *Appl. Opt.* **2017**, *56*, 9888–9891. [CrossRef]
40. Von Freymann, G.; Kitaev, V.; Lotsch, B.V.; Ozin, G.A. Bottom-up assembly of photonic crystals. *Chem. Soc. Rev.* **2013**, *42*, 2528–2554. [CrossRef]
41. Inoue, T.; Morita, R.; Nigo, K.; Yoshida, M.; De Zoysa, M.; Ishizaki, K.; Noda, S. Self-evolving photonic crystals for ultrafast photonics. *Nat. Commun.* **2023**, *14*, 50. [CrossRef]
42. Li, H.; Wu, P.; Zhao, G.; Guo, J.; Wang, C. Fabrication of industrial-level polymer photonic crystal films at ambient temperature Based on uniform core/shell colloidal particles. *J. Colloid Interface Sci.* **2021**, *584*, 145–153. [CrossRef] [PubMed]

crystals

MDPI

Article

Embedded Rough-Neck Helmholtz Resonator Low-Frequency Acoustic Attenuator

Xianming Sun [1], Tao Yu [2,*], Lipeng Wang [1], Yunshu Lu [1] and Changzheng Chen [3]

1 School of Mechanical and Automotive Engineering, Ningbo University of Technology, Ningbo 315211, China; sxm_nbkb@163.com (X.S.); wlp_syuct@163.com (L.W.); luyunshu2013@sina.com (Y.L.)
2 School of Environmental and Chemical Engineering, Shenyang University of Technology, Shenyang 110870, China
3 School of Mechanical Engineering, Shenyang University of Technology, Shenyang 110870, China; czchen@sut.edu.cn
* Correspondence: yutao@smail.sut.edu.cn; Tel.: +86-150-0962-9231

Abstract: In various practical noise control scenarios, such as duct noise mitigation, industrial machinery, architectural acoustics, and underwater applications, it is essential to develop noise absorbers that deliver effective low-frequency attenuation while maintaining compact dimensions. To achieve low-frequency absorption within a limited spatial volume, this study proposes an embedded Helmholtz resonator featuring a roughened neck and establishes a numerical computational model that incorporates thermos viscous effects. A quantitative investigation is conducted on three types of embedded rough-neck geometries (rectangular-grooved, triangular-grooved, and undulated) to elucidate their acoustic performance, with particular attention to differences in acoustic transmission loss and acoustic impedance characteristics. In response to the practical demand for even lower-frequency attenuation, this work further focuses on optimizing the structural parameters of an embedded rectangular-grooved Helmholtz resonator (ERHR). A back-propagation (BP) neural network models and predicts how structural parameters impact the acoustic transmission coefficient, elucidating the effects of geometric variations. Moreover, by coupling the BP network with the Golden Jackal Optimization (GJO) algorithm, a BP-GJO optimization model is developed to refine the structural parameters. The findings reveal that the proposed method significantly improves resonator spatial utilization at a specific noise frequency while preserving acoustic transmission loss performance. This work thereby provides a promising strategy for designing low-frequency, compact Helmholtz resonators suitable for a wide range of noise control applications.

Keywords: acoustic metamaterials; Helmholtz resonator; parameter optimization; pipeline noise

1. Introduction

Low-frequency noise is a critical topic in noise control research. Acoustic metamaterials show great potential due to their unique acoustic manipulation capabilities beyond conventional materials [1]. These materials can be categorized into phononic crystals and locally resonant metamaterials. Phononic crystals rely on periodic structures to create bandgaps through Bragg scattering, effective at wavelengths comparable to the lattice constant, while locally resonant metamaterials achieve unique properties through subwavelength resonances, with effective medium theories explaining phenomena such as negative mass density [2,3] and negative refractive index [4]. In recent years, various acoustic metamaterial structures have been developed, including locally resonant phononic crystals [5,6], Helmholtz resonators [7–10], membrane-based acoustic metamaterials [11–13], and duct-type silencers [14,15]. However, one major challenge in practical applications of acoustic metamaterials is that achieving low-frequency noise attenuation often requires large structural dimensions, which significantly restricts their feasibility and application potential in engineering contexts. Therefore, further exploration of acoustic metamaterials with high spatial efficiency and broadband low-frequency sound insulation is essential.

Helmholtz resonators play a significant role in sound absorption and noise reduction applications. Comprising a closed cavity and a narrow neck, these resonators typically have a geometric scale much smaller than the sound wavelength. Fu [16] proposed a folded Helmholtz resonator and demonstrated that, for the same resonator volume, the folded design achieves a lower effective frequency and a transmission loss of up to 35 dB. Wu [17] introduced a rotating Helmholtz resonator as part of a phononic crystal duct, achieving a total bandgap width of 2297 Hz and a transmission loss exceeding 10 dB, effectively mitigating broadband noise from air compressors. Hedayati [18] tested a passive metamaterial comprising 64 Helmholtz resonator units, achieving 18 dB and 33 dB sound attenuation at 150 Hz and 350 Hz in a closed system and 15–20 dB attenuation in an open system. Li [19] designed an array of Helmholtz resonators for broadband sound absorption, demonstrating a continuous absorption capability in the 300–2200 Hz range. Chen proposed a Helmholtz resonator with side slits, achieving absorption coefficients exceeding 80% in the 470–930 Hz range by varying resonator heights across different units. Kong [20] designed an arc-shaped Helmholtz resonator with a negative Poisson's ratio for underwater low-frequency sound absorption, significantly lowering the peak resonance frequency and achieving an average absorption coefficient exceeding 0.9 within 520–930 Hz, with a peak value of 0.99 at 740 Hz. Notably, the resonator's thickness remained in the ultra-subwavelength regime. Yang [21] designed an embedded spiral Helmholtz resonator acoustic metamaterial, enhancing system impedance tunability and reducing material thickness, achieving nearly perfect absorption in the 102–164 Hz range.

In the field of duct noise control, Helmholtz resonators exhibit superior acoustic performance in specific frequency ranges, making them a core technology in silencer design [22]. Yang [23] developed a rectangular duct silencer incorporating Helmholtz resonators, achieving a transmission loss of 17.3 dB within a specific frequency range, significantly contributing to noise reduction in Roots blowers. Augusto [24] designed a labyrinthine Helmholtz resonator for duct noise control, achieving 28 dB sound attenuation near 182 Hz, with a total thickness 20 times smaller than the working wavelength. Han [25] designed a dual-Helmholtz resonator with an integrated soundproofing barrier, achieving ventilation and noise isolation, with two effective attenuation bands near the first Fabry–Pérot resonance frequency. Zhang [15] developed a subwavelength sector acoustic metamaterial for duct noise attenuation, quantitatively revealing the effects of structural parameters on bandwidth and peak frequency. The study showed that thermal dissipation occurs within the cavity, while viscous dissipation is mainly present in the ducts, achieving a noise attenuation of 10 dB in the 700–1100 Hz range. Cao [26] investigated the acoustic attenuation of side-branched Helmholtz resonators with embedded apertures in grazing flow ducts, achieving excellent noise reduction performance by extending the embedded aperture length. Yaw [27] introduced a sliced Helmholtz resonator structure with roughened necks for subwavelength low-frequency sound absorption, achieving an average absorption coefficient exceeding 0.9 in the 50–170 Hz range.

In practical applications, spatial constraints impose strict requirements on minimizing the volume of acoustic materials. To optimize acoustic performance, it is necessary to carefully design the geometric parameters and structural layout of acoustic materials. Lv [28] proposed an iterative optimization method based on a genetic algorithm to optimize Helmholtz resonator shapes, significantly enhancing low-frequency filtering performance without changing the volume. The optimized resonator exhibited a 10% reduction in resonance frequency, with effective low-frequency filtering in the 84–232 Hz range. Gu [29] designed an acoustic liner structure for low-frequency broadband noise absorption by coupling Helmholtz resonators with extended necks, using a particle swarm algorithm to identify optimal parameters, achieving absorption coefficients exceeding 80% in the 800–1000 Hz range, effectively suppressing large bypass turbofan engine noise. Jiao [30] proposed a multi-layer microperforated plate absorber supported by Helmholtz resonators, achieving absorption coefficients above 0.8 in the 296–1000 Hz range through genetic algorithm optimization. Bi [31] investigated the sound absorption performance of Helmholtz resonators with spatial segmentation and chamber grouping, obtaining optimal geometric parameters using finite element simulations and the cuckoo search algorithm, achieving absorption coefficients above 0.8 in the 400–800 Hz range.

As the sound wave enters the Helmholtz resonator, the sound wave interacts with the cavity and neck structure, creating a resonance that dissipates energy through viscous and thermal losses. In the design of Helmholtz resonators, the geometric dimensions of the neck significantly influence the sound absorption frequency [32]. However, although conventional metasurface-based Helmholtz resonators are effective at certain discrete frequencies, achieving substantial attenuation and enhanced acoustic filtering performance at low frequencies remains challenging. Due to inherent spatial constraints, simply increasing the neck's geometric dimensions inevitably leads to an enlargement of the overall structure. To address this limitation, this study proposes an embedded rough-neck design that extends the effective neck length within a limited cavity space, thereby achieving lower sound absorption frequencies and superior acoustic filtering performance. This improvement provides a novel approach for optimizing the acoustic performance of Helmholtz resonators. The structure of this paper is organized as follows. Section 1 introduces the existing designs. Section 2 establishes the numerical simulations incorporating thermos-viscous effects. Section 3 presents experimental validation and comparative analysis of acoustic performance. Section 4 applies a BP neural network and GJO algorithm for structural optimization. Finally, Section 5 concludes the study with a summary of key findings.

2. Theoretical Foundation and Finite Element Model

The geometry of the embedded neck plays a crucial role in tuning the resonance frequency and controlling the impedance of the resonator. In order to explore the effects of different neck geometries on the acoustic performance of the resonator, this paper proposes three types of embedded rough-necked Helmholtz resonators: rectangular-groove neck, triangular-groove neck, and wave neck. The schematic diagram of the proposed acoustic model consists of three embedded rough-neck Helmholtz resonators, as shown in Figure 1a. Figure 1b shows the embedded rectangular-groove neck, Figure 1c shows the embedded triangular-groove neck, and Figure 1d shows the embedded wavy neck. In this design, R represents the pipe radius, t represents the thickness of the cavity and embedded neck, h represents the width of the embedded neck, H represents the cavity height, R_B represents the cavity radius, and l represents the length of the embedded neck. The harmonic plane wave enters from the top, represented by p_{in}, and exits from the bottom, represented by p_{out}.

When resonators are arranged in series, their coupling effect results in a broader absorption frequency range, enabling broadband low-frequency absorption. To further save space, as shown in Figure 1a, adjacent units share a common cavity wall. To systematically study the acoustic performance of the embedded rough-neck in individual and series-arranged Helmholtz resonators, a numerical model of this structure was developed using commercial software COMSOL Multiphysics 6.2, as shown in Figure 2a. The model includes Perfectly Matched Layers (PMLs), a background pressure field, and Helmholtz resonators, which are composed of rough necks and cavities. The pressure acoustics module is applied to the resonator cavity and external incident field regions, treated as non-damping regions, while the thermo-viscous acoustics module is applied to the rough neck of the resonator. As shown in Figure 2b, the PMLs and background pressure field are meshed using mapped meshing, while free tetrahedral mesh elements are used for the apertures and Helmholtz resonators, with boundary layer adjustments made to improve simulation accuracy in boundary regions. The necessary physical parameters of air are listed in Table 1. The background pressure field has an acoustic pressure amplitude of 1 Pa, and its pressure distribution is governed by the Helmholtz equation:

$$\nabla \cdot (-\frac{1}{\rho_0}\nabla p) - \frac{\omega^2}{\rho_0 {c_0}^2} p = 0 \quad (1)$$

where p is the total sound pressure, ω is the angular frequency, ρ_0 is the air density, and c_0 is the speed of sound.

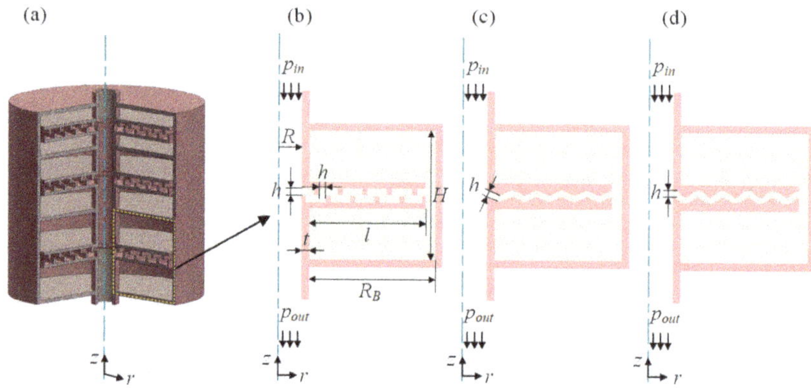

Figure 1. Embedded rough-neck Helmholtz resonator model (**a**) 3D model; (**b**) Rectangular-groove neck; (**c**) Triangular-groove neck; and (**d**) wavy neck.

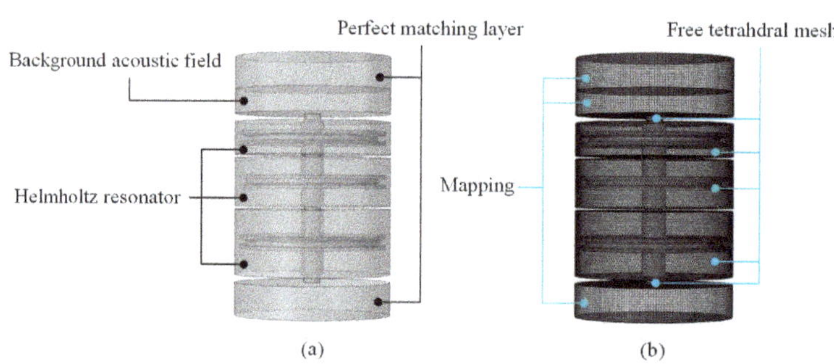

Figure 2. (**a**) Finite element model; (**b**) Finite element mesh division.

Table 1. Physical parameters of air.

Parameters	Value
Bulk viscosity (Pa·s)	1.11×10^{-5}
Thermal conductivity ($W \cdot m^{-1} \cdot K^{-1}$)	2.63×10^{-2}
Density ($kg \cdot m^{-3}$)	1.21
Ratio of specific heat	1.40
Heat capacity at constant pressure ($J \cdot kg^{-1} \cdot s^{-1}$)	1005.71
Dynamic viscosity (Pa·s)	1.85×10^{-5}
Speed of sound ($m \cdot s^{-1}$)	343.00

3. Validation and Analysis

This study used a four-pole acoustic testing system to experimentally investigate the sound transmission loss under dual-load conditions. Compared to the three-pole method, the four-pole method reduces experimental complexity [33]. An impedance tube kit type SW422 was used according to the international standards ISO 10534-2 [34] and ASTM E2611-12 [35]. The schematic diagram of the experimental setup is shown in Figure 3, where the acoustic signals are transmitted from the power amplifier (Type PA50) to the sound source, while the signals from four 1/4 inch condenser microphones are captured through a signal collector (Type MC3242), and the test results are analyzed by the software system. The power amplifier has a frequency response range of 20 Hz–20 kHz, harmonic distortion of less than 0.05%, and a S/N of more than 95 dB. The sensitivity of the sound source is 89.8 dB. The impedance tube, with a diameter of 100 mm, is equipped with four microphones, designated as m_1, m_2, m_3, and m_4. The distance between m_1 and m_2 is s_1 = 50 mm, while the sample is positioned l_1 = 60 mm and l_2 = 60 mm from m_2 and m_3, respectively. The distance between m_3 and m_4 is s_2 = 50 mm. The diameter of the sample is $D = 2(R_B + 2t + R)$, and the height is t. The front surface of the sample is defined as the first reference plane ($x = 0$), and the second reference plane ($x = t$) is located at a distance t from the first reference plane. Dual-load conditions are achieved by changing the impedance at the second waveguide end, Z_i (i = m, n). The rotational embedded rough-neck Helmholtz resonator samples were fabricated via stereolithography, with a 0.05 mm layer height, 0.2 mm nozzle diameter, and ±0.01 mm repeat positioning precision. The formula for calculating the transmission loss is

$$\begin{cases} \mathrm{STL} = 20\log_{10}\left|\frac{1}{\tau}\right| \\ \tau = \frac{T_{11}+\frac{T_{12}}{Z_0}+T_{21}Z_0+T_{22}}{2e^{jkd}} \end{cases} \quad (2)$$

where τ is the experimentally determined transmission coefficient, and T represents the transmission matrix obtained from measurements [36], while Z_0 is the characteristic impedance of air.

To further study the acoustic performance of the model, a comparative analysis of the acoustic characteristics of three different types of embedded rough-neck Helmholtz resonators was conducted. This includes resonators with an embedded rectangular-groove neck, an embedded triangular-groove neck, and an embedded wavy neck. Their structural parameters are listed in Table 2. The obtained sound transmission loss curves are shown in Figure 4a, with the STL peaks for the three resonators appearing at 396 Hz, 419 Hz, and 426 Hz, corresponding to STL values of 50.86 dB, 51.11 dB, and 51.20 dB, respectively. Notably, compared to the embedded triangular-groove neck and embedded wavy neck resonators, the embedded rectangular-groove neck resonator achieved a lower noise reduction frequency. The transmission coefficient curves are shown in Figure 4b, with the trough frequencies of the transmission coefficient corresponding to the peak frequencies of the sound transmission loss. Sound waves are reflected by near-field radiation at the side branch entry, causing attenuation of the incident waves, which results in reduced transmission coefficients. In addition, resonance within the channel significantly enhances energy conversion near the wall, further reducing the transmission coefficient. Under conditions where the transmission coefficient is less than 0.4, the effective bandwidth of the embedded triangular-groove neck and embedded wavy neck resonators is significantly larger than that of the other two resonator structures, and they also demonstrate good sound attenuation at higher frequencies (e.g., 1000 Hz). This phenomenon can be attributed to impedance matching and the acoustic siphon effect [37], although their STL peak frequencies are higher.

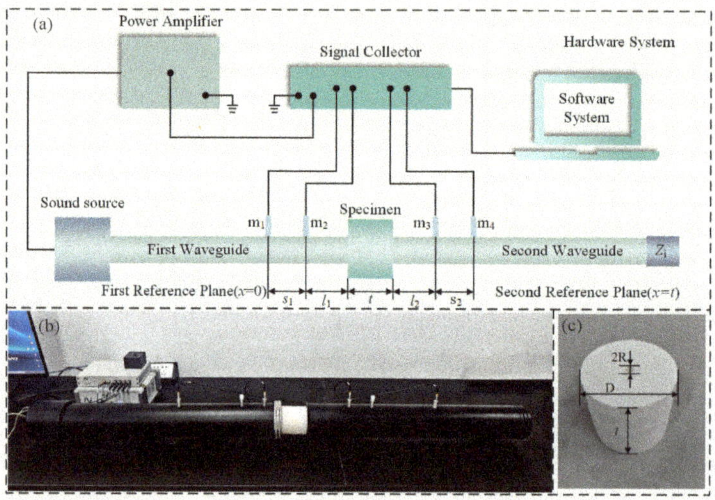

Figure 3. Components of the test system and samples of the embedded rough-neck Helmholtz resonators. (**a**) Schematic of the experimental system layout; (**b**) Experimental apparatus; and (**c**) resonator samples.

Table 2. Geometric parameter (in mm).

R	t	h	H	R_B	l
5	2	2.5	40	41	38

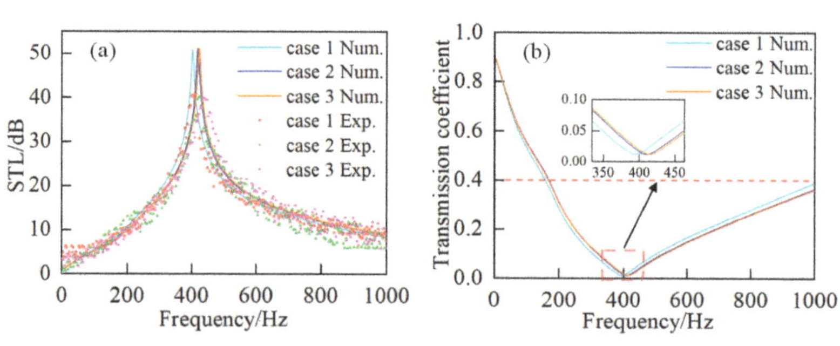

Figure 4. (**a**) Sound transmission loss; (**b**) Transmission coefficient. (case 1 indicates rectangular-grooved; case 2 indicates triangular-grooved; and case 3 indicates undulated).

The impedance of the resonator and how well it matches the impedance of the surrounding medium (air) significantly influences its acoustic performance. When the impedance Z equals the characteristic impedance of the medium $\rho_0 c_0$, the sound absorption efficiency is maximized, where the reflection coefficient is 0, the transmission coefficient is 0, and the absorption coefficient is 1, resulting in perfect absorption [38]. To analyze the model's noise reduction mechanism, the acoustic impedance at the entry of three resonator types was evaluated, as shown in Figure 5. At resonance, the impedance $|Z_n|$ values were 1.15 Pa·s/m, 0.147 Pa·s/m, and 0.136 Pa·s/m for the rectangular-groove, triangular-groove, and wavy necks, respectively. The real part of the acoustic impedance $\mathrm{Re}(Z_n) > 0$, indicates the presence of sound energy dissipation. The imaginary part $\mathrm{Im}(Z_n)$, representing the acoustic reactance, exhibits an increasing trend with frequency. Meanwhile, the minimum value of the impedance modulus $|Z_n|$ corresponds to the frequency of the transmission coefficient trough. However, due to thermo-viscous losses, the minimum value of $|Z|$ is not zero, indicating that the impedance at the side branch entry is not fully matched. This imperfect impedance matching leads to some degree of wave reflection and transmission while simultaneously enhancing the multiple scattering and dissipation of sound waves within the structure, thereby improving noise reduction performance. The sound pressure level distributions for the embedded rectangular-groove neck, embedded triangular-groove neck, and embedded wavy neck resonators at the resonant frequency are shown in Figure 5d–f. Due to resonance in the double spiral channels, a maximum sound pressure level appears inside, indicating satisfactory noise reduction performance for the double spiral resonator.

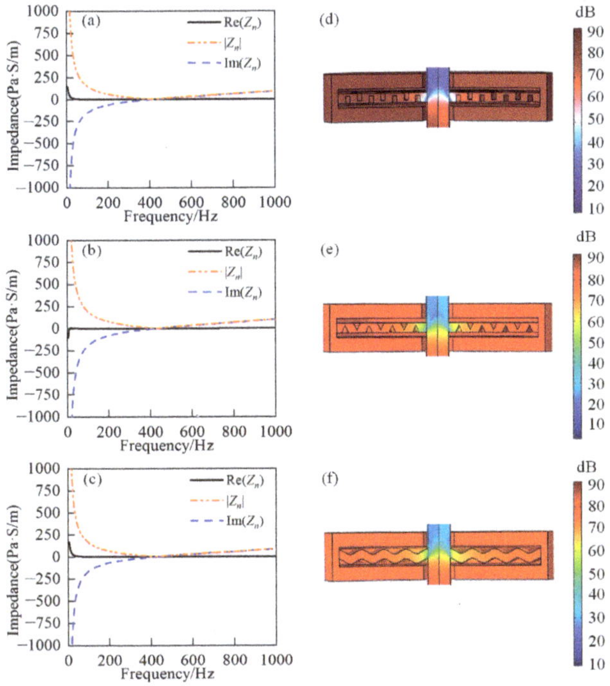

Figure 5. Resonator inlet impedance (**a**) Rectangular-grooved; (**b**) Triangular-grooved; (**c**) Undulated; and structural sound pressure level distribution; (**d**) Rectangular-grooved; (**e**) Triangular-grooved; (**f**) Undulated.

4. Prediction and Optimization

4.1. Structural Parameter Analysis Using BP Network

To achieve lower sound absorption frequencies, this study analyzed the parameters of the embedded rectangular-groove neck Helmholtz resonator. Based on the complex nonlinear relationship between the structural dimensions of the resonator and the transmission coefficients and the corresponding frequencies, the transmission coefficients of the resonators are predicted using a BP network, which is highly adaptive in predicting acoustic performance metrics. The BP network model is more advantageous in expressing the nonlinear relationship than the orthogonal experiment and the response surface method, and the accuracy of its outputs is crucial as the key constraint [39]. In the prediction model, the acoustic performance of the rotational embedded rectangular-groove neck Helmholtz resonator is primarily affected by four structural parameters: the neck length l, cavity height H, radius of the cavity R_B, and the number of rectangular grooves T. The BP network uses structural parameters as inputs and frequency with sound transmission loss as outputs. It comprises four input nodes, two output nodes, and ten hidden layer nodes, with the input nodes reflecting parameters affecting acoustic performance. Training data, derived from 1000 subsamples, were divided as 80% for training, 10% for validation, and 10% for testing. The training process effectively established the relationship between structural parameters and acoustic performance.

$$\begin{cases} \mathrm{STL} = f_1(l, H, R_B, T) \\ f_s = f_2(l, H, R_B, T) \end{cases} \tag{3}$$

where STL represents sound transmission loss, and f_s represents the frequency corresponding to STL.

The regression analysis results are shown in Figure 6a. A comparison of predicted and actual values across 1000 parameter combinations in the test set shows strong regression performance, achieving a maximum fitting coefficient of 0.996. Further analysis of prediction errors for the training, validation, and test sets shows that the error distribution follows a normal distribution with very small error values in Figure 6b; this confirms that the model's prediction accuracy aligns with expectations. Figure 7 presents the mean square error (MSE) function for training, validation, and test sets. The training process quickly converged within 20 iterations, with MSE values stabilizing after 1000 iterations. The final total number of training iterations was 1005. The MSE for the training, validation, and test sets are 4.19×10^{-2}, 3.37×10^{-2} and 4.17×10^{-2}, respectively.

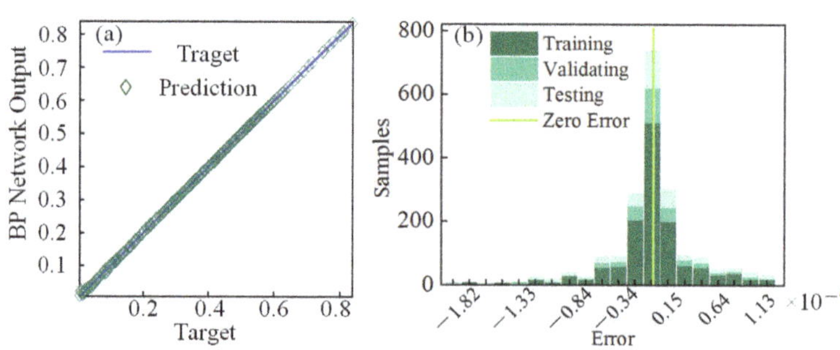

Figure 6. BP network prediction performance. (**a**) Regression results; (**b**) Error distribution analysis.

The influence of structural parameters on the transmission coefficient is shown in Figure 8, where the color change corresponds to variations in the transmission coefficient value. Four parameters affecting the transmission coefficient and corresponding frequency were studied: the length of the embedded neck l, the cavity height H, the number of rectangular grooves T, and the effective radius of the cavity R_B. When H = 40 mm, T = 7, and R_B = 41 mm, the variation of the transmission coefficient is shown in Figure 8a. As the embedded neck length increases, the transmission coefficient trough value increases, and the corresponding resonance frequency shifts towards lower frequencies. When l = 38 mm, T = 7, and R_B = 41 mm, the variation of the transmission coefficient is shown in Figure 8b. As the cavity height increases, the transmission coefficient trough value slightly increases, and the corresponding resonance frequency decreases. When l = 38 mm, H = 40 mm, and R_B = 41 mm, the variation of the transmission coefficient is shown in Figure 8c. As the number of rectangular grooves increases, the transmission coefficient trough value decreases, and the corresponding resonance frequency decreases, along with a reduction in bandwidth. When l = 38 mm, H = 40 mm, and T = 7, the variation of the transmission coefficient is shown in Figure 8d. As the effective radius of the cavity increases, the transmission coefficient trough value tends to increase, and the corresponding resonance frequency decreases.

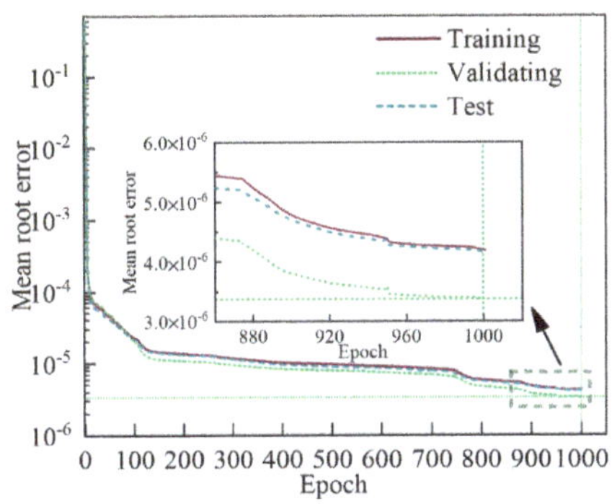

Figure 7. Mean Square Error diagram.

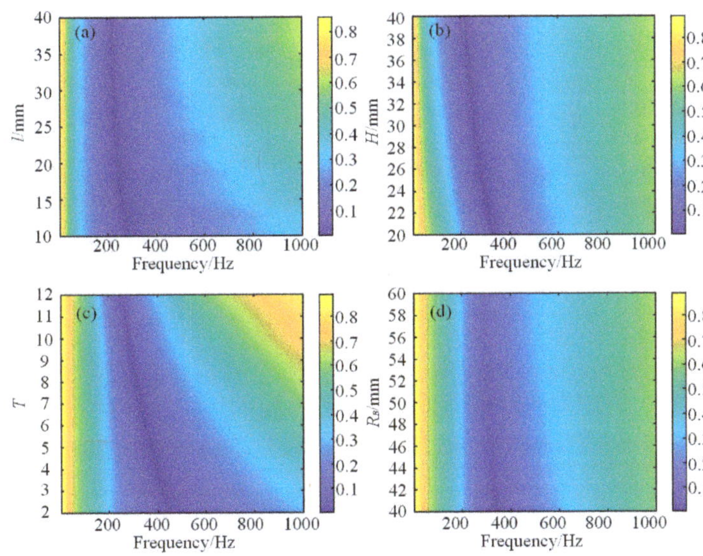

Figure 8. Variation of transmission coefficient with structural parameters. (**a**) Neck length l; (**b**) Cavity height H; (**c**) Number of rectangular grooves T; (**d**) Effective radius of cavity R_B.

The overall framework of the proposed optimization model is shown in Figure 9. The model is divided into two stages. In the first stage, a BP network is utilized to build a nonlinear mapping between the resonator's structural parameters and its acoustic performance. X denotes the input parameters, i.e., the resonator structural parameter dataset, which are used as features for model training. Y denotes the output parameters, i.e., the acoustic transmission loss. '*' denotes the optimised data. Hidden layer nodes are used to capture the complex non-linear relationship between X (X^*) and Y (Y^*). Training the BP network captures the intricate relationship between structural parameters and desired acoustic performance, offering precise constraints for optimization. The second stage combines the BP network with the GJO algorithm. Based on the constraints obtained in the first stage, the goal is to minimize the resonator volume while ensuring good acoustic performance.

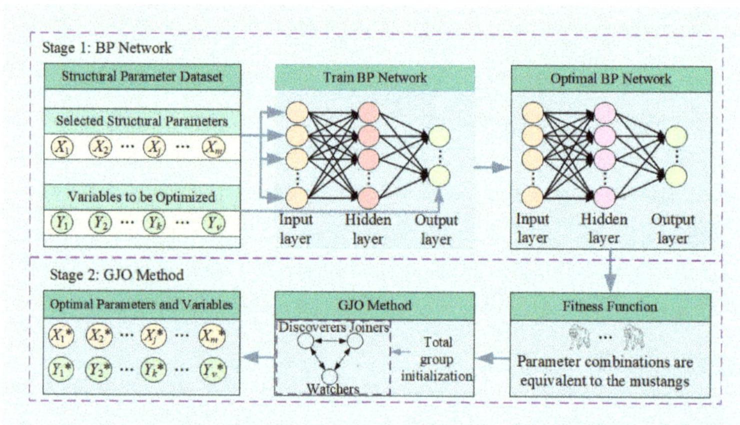

Figure 9. Optimization algorithm framework.

The acoustic optimization goal is to minimize the resonator volume Γ while using sound transmission loss and the corresponding frequency as equation constraints. Assuming the actual sound transmission loss is Q_1 and the desired transmission loss exceeds Q_2, the value ranges for the structural parameters, shown in Table 3, are defined to enhance the quality of the initial solution and ensure diversity in the parameter search. The fitness function is formulated in Equation (4), with equality and inequality constraints described in Equation (5). When constraints are not satisfied, a penalty term $o = o_1 + \beta \sum_i c_{eq}(x_i)^2$ is introduced, $\beta = 2000$ is the penalty factor to guide the algorithm toward finding feasible solutions.

$$o_1 = \min \Gamma(l, H, R_B, T) \tag{4}$$

Table 3. Optimize the range of parameter constraints.

ub_1/(mm)	ub_2/(mm)	ub_3/(mm)	ub_4(mm)	lb_1/(mm)	lb_2/(mm)	lb_3/(mm)	lb_4
40	40	60	2	20	20	41	12

$$\begin{cases} f_s = Q_1 \\ lb_1 \leq l \leq ub_1 \\ lb_2 \leq H \leq ub_2 \\ lb_3 \leq R_B \leq ub_3 \\ lb_4 \leq T \leq ub_4 \\ \text{STL} \geq Q_2 \end{cases} \tag{5}$$

The GJO algorithm uses a population search method to initialize the optimization variables as uniformly distributed within the search space, with the initial solution range determined by the upper $X_{\max}$ and lower $X_{\min}$ bounds of the variables

$$X_0 = X_{\min} + rand(X_{\max} - X_{\min}) \tag{6}$$

where *rand* is a random number between 0 and 1.

The initial matrix *Prey* is created during initialization. The first and second adapters are the primary and auxiliary design parameters. The matrix *Prey* is represented as

$$Prey = \begin{bmatrix} X_{1,1} & X_{1,2} & \cdots & X_{1,d} \\ X_{2,1} & X_{2,1} & \cdots & X_{2,d} \\ \vdots & \vdots & \ddots & \vdots \\ X_{n,1} & X_{n,2} & \cdots & X_{n,d} \end{bmatrix} \tag{7}$$

where $X_{i,j}$ represents the *i*-th solution's *j*th dimension. There are n solutions and *d* variables. During optimization, a fitness function is constructed to evaluate the fitness of each solution, and the subsequent matrix collects all solution fitness values

$$F_{OA} = \begin{bmatrix} f(X_{1,1}; X_{1,2}; \cdots; X_{1,d}) \\ f(X_{2,1}; X_{2,2}; \cdots; X_{2,d}) \\ \vdots \\ f(X_{n,1}; X_{n,2}; \cdots; X_{n,d}) \end{bmatrix} \tag{8}$$

where F_{OA} is the matrix that preserves the fitness of each solution, and f is the objective function. The two solutions with the best performance in the system are designated as the primary and secondary solutions, while other parameters are dynamically adjusted based on these.

The position update formulas for the primary and secondary solutions are

$$X_1(t) = X_M(t) - E.|X_M(t) - rl.Prey(t)| \tag{9}$$

$$X_{2'}(t) = X_{FM}(t) - E.|X_{FM}(t) - rl.Prey(t)| \tag{10}$$

where t represents the current iteration, *Prey(t)* is the solution position vector, and $X_M(t)$ and $X_{FM}(t)$ are the positions of the primary and secondary solutions, respectively. $X_1(t)$ and $X_2(t)$ are the latest positions of the primary and secondary solutions. $E = E_1 E_0$ represents the escape energy, $E_1 = c_1(1 - (t/T))$ indicating a decrease in solution energy, $E_0 = 2r - 1$ denotes the initial state of the solution's energy, and r is any number between 0 and 1. T represents the maximum number of iterations, c_1 is a constant equal to 1.5, and rl is a random number vector based on the Levy distribution, representing Levy flight [40].

After the update, the current global optimal solution's position is represented as

$$X(t+1) = \frac{X_1(t) + X_2(t)}{2} \tag{11}$$

4.2. Optimization Results

The optimization results obtained using the Golden Jackal Optimization algorithm and the corresponding structural parameters are listed in Table 4. Compared with the original structural parameters, the adjusted optimization model significantly reduced the overall volume of the resonator. To further verify the optimization results, a numerical model was built with the optimized structural parameters. The changes in sound transmission loss and transmission coefficient before and after optimization are shown in Figure 10. The optimized structure improved spatial efficiency while maintaining excellent noise reduction performance, especially in the 200–750 Hz frequency range, where the sound transmission loss remained at 30 dB. Compared with the original structure, the optimized design not only effectively reduced the resonator volume but also maintained ideal acoustic performance in key frequency bands, demonstrating that the optimization algorithm successfully improved spatial efficiency without negatively impacting noise reduction.

Table 4. Optimised geometric parameter (in mm).

	L/(mm)	H/(mm)	T	R_B/(mm)
First layer	41	21	4	43
Second layer	41	30	7	43
Third layer	41	35	12	43

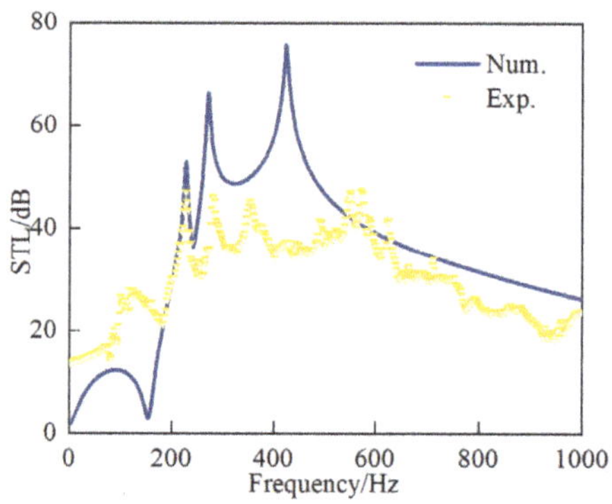

Figure 10. Optimized sound transmission loss of ERHRs.

5. Conclusions

This study proposes a novel acoustic structure based on an embedded rectangular-groove neck Helmholtz resonator. The corresponding numerical model was established, and its acoustic properties were analyzed, comparing the acoustic performance of different rough-neck designs. The model was experimentally validated. The BP-GJO optimization model was constructed to optimize the spatial efficiency of the embedded rectangular-groove neck Helmholtz resonator at targeted frequencies. The primary conclusions are as follows:

Theoretical and experimental results verified the accuracy of the theoretical model. At the resonant frequency, the embedded rough-neck Helmholtz resonator exhibited imperfect impedance matching due to thermo-viscous losses.

The BP neural network was employed to model and predict the correlation between structural parameters and the acoustic performance of the embedded rectangular-groove neck Helmholtz resonator, providing insights into the influence of structural parameters on the resonator's transmission coefficient.

The BP-GJO optimization model effectively optimized the structural parameter combinations, significantly enhancing spatial efficiency without compromising the acoustic performance of the embedded rectangular-groove Helmholtz resonator.

Author Contributions: Conceptualization, X.S., T.Y. and L.W.; methodology, X.S. and T.Y.; software, X.S. and T.Y.; validation, T.Y. and X.S.; formal analysis, X.S., T.Y., and C.C.; investigation, X.S.; resources, L.W. and Y.L.; data curation, X.S. and L.W.; writing—original draft preparation, X.S.; writing—review and editing, X.S.; visualization, Y.L. and T.Y.; supervision, L.W.; project administration, L.W., Y.L. and C.C.; funding acquisition, X.S. and L.W. All authors have read and agreed to the published version of the manuscript.

Funding: This work was supported by the State Key Laboratory for Joint Open Fund (2022-KF-22-11), the State Key Laboratory for Joint Open Fund (2022-KF-24-04), and the Ningbo River Talent Attraction Project for Innovative Talents in Science and Technology Services in the Field of Urban Economy (2021B-034-G).

Data Availability Statement: The detailed data supporting the results of this study are available from the corresponding authors upon request.

Conflicts of Interest: The authors declare no conflicts of interest.

References

1. Aydin, G.; San, S.E. Breaking the limits of acoustic science: A review of acoustic metamaterials. *Mater. Sci. Eng. B* **2024**, *305*, 117384. [CrossRef]
2. Liu, Z.; Zhang, X.; Mao, Y.; Zhu, Y.Y.; Yang, Z.; Chan, C.T.; Sheng, P. Locally Resonant Sonic Materials. *Science* **2000**, *289*, 1734–1736. [CrossRef] [PubMed]
3. Huang, H.; Sun, C.; Huang, G. On the negative effective mass density in acoustic metamaterials. *Int. J. Eng. Sci.* **2008**, *47*, 610–617. [CrossRef]
4. Smith, R.D.; Pendry, B.J.; Wiltshire, K.C.M. Metamaterials and Negative Refractive Index. *Science* **2004**, *305*, 788–792. [CrossRef] [PubMed]
5. Chai, Z.; Liu, H.; Xiang, J. Low-frequency broadband vibration reduction based on a square spiral beam local resonance phononic crystal. *Jpn. J. Appl. Phys.* **2024**, *63*, 034002. [CrossRef]
6. Sun, X.-W.; Xu, G.-G.; Li, R.-S.; Tan, M.-T.; Gao, X.-L.; Sun, W.-B. Reconfigurable local-resonance elastic waveguides in piezoelectric phononic crystals plate. *J. Intell. Mater. Syst. Struct.* **2024**, *35*, 750–759. [CrossRef]
7. El Malki, M.; Khettabi, A.; Sallah, M.; Zaky, Z.A. Noise filter using a periodic system of dual Helmholtz resonators. *Sci. Rep.* **2024**, *14*, 24987. [CrossRef]
8. Papadakis, N.M.; Stavroulakis, G.E. FEM Investigation of a Multi-neck Helmholtz Resonator. *J. Vib. Eng. Technol.* **2024**, *13*, 10610. [CrossRef]
9. Zhang, L.; Xin, F. Perfect low-frequency sound absorption of rough neck embedded Helmholtz resonators. *J. Acoust. Soc. Am.* **2022**, *151*, 1191–1199. [CrossRef] [PubMed]
10. Zhu, J.; Qu, Y.; Gao, H.; Meng, G. Nonlinear sound absorption of Helmholtz resonators with serrated necks under high-amplitude sound wave excitation. *J. Sound Vib.* **2022**, *537*, 117197. [CrossRef]
11. Xu, H.; Kong, D. A thin-film acoustic metamaterial absorber with tunable sound absorption characteristics. *J. Acoust. Soc. Am.* **2023**, *153*, 3493–3500. [CrossRef]
12. Zhang, D.; Su, X.; Sun, Y.; Chen, C.; Sun, X. Mechanism Analysis and Experiment Study for Wire Mesh-Assisted Ventilated Acoustic Metamaterials Based on the Acoustic Analytical Model and Numerical Acoustic-Flow Coupling Model. *J. Vib. Eng. Technol.* **2024**, *12*, 6649–6663. [CrossRef]
13. Wang, L.B.; Wu, J.H.; Lei, Y.Z.; Niu, J.M.; Huang, Y.; Liu, C.R. A novel membrane-cavity-grating (MCG) meta-structure for enhancing low-frequency sound absorption. *J. Phys. D Appl. Phys.* **2022**, *55*, 395502. [CrossRef]
14. Zhang, D.; Su, X.; Sun, Y.; Luo, Y.; Sun, X.; Chen, C. Performance study and improvement of space-folded metamaterial muffler for pipe under grazing flow. *Appl. Acoust.* **2024**, *220*, 109984. [CrossRef]
15. Zhang, D.; Su, X.; Sun, Y.; Chen, C.; Sun, X. Numerical simulation and experimental study of a broadband acoustic metamaterial duct muffler considering thermal-viscous loss. *J. Mech. Sci. Technol.* **2024**, *38*, 1039–1049. [CrossRef]
16. Fu, T.; Chen, C.; Zhang, D.; Sun, X. Numerical Calculation and Experimental Study of Low Frequency Helmholtz Resonators with Folded Features. *Noise Vib. Control* **2024**, *44*, 289–294.

17. Wu, X.; Liu, M.; Guan, H. Collaborative Optimization of Phononic Crystal Pipe Based on Revolving Helmholtz Resonators for Air Compressor System in FCEVs. *J. Vib. Eng. Technol.* **2024**, *12*, 7035–7044. [CrossRef]
18. Hedayati, R.; Lakshmanan, S.P. Active Acoustic Metamaterial Based on Helmholtz Resonators to Absorb Broadband Low-Frequency Noise. *Materials* **2024**, *17*, 962. [CrossRef]
19. Li, X.; Mao, Q. A Helmholtz Resonator Array for Low-Frequency Broadband Sound Absorption. In Proceedings of the 6th China Aeronautical Science and Technology Conference, Wuzhen, China, 20 December 2023.
20. Kong, W.; Fu, T.; Rabczuk, T. Improvement of broadband low-frequency sound absorption and energy absorbing of arched curve Helmholtz resonator with negative Poisson's ratio. *Appl. Acoust.* **2024**, *221*, 110011. [CrossRef]
21. Yang, M.; Huang, X.; Jiao, W.; Song, J.; Chen, M.; Xie, X. Tunable Low-frequency Sound Absorber via Helmholtz Resonators with Embedded Spiral Tube. In Proceedings of the 2024 IEEE International Conference on Mechatronics and Automation (ICMA), Tianjin, China, 4–7 August 2024.
22. Gao, C.; Hu, C.; Hou, B.; Zhang, X.; Li, S.; Wen, W. Ventilation duct silencer design for broad low-frequency sound absorption. *Appl. Acoust.* **2023**, *206*, 109324. [CrossRef]
23. Yang, S.; Lee, J.; Yu, J. Internal structure optimization for noise reduction in next-generation blower silencers. *J. Mech. Sci. Technol.* **2024**, *38*, 2223–2230. [CrossRef]
24. Beck, A.B.; Almeida, G.d.N.; Mikulski, R.Z.; Vergara, E.F. Low-frequency acoustic attenuator based on a labyrinthine Helmholtz resonator. *J. Braz. Soc. Mech. Sci. Eng.* **2024**, *46*, 407. [CrossRef]
25. Han, I.; Lee, I.; Yoon, G. Studies on Dual Helmholtz Resonators and Asymmetric Waveguides for Ventilated Soundproofing. *Sensors* **2024**, *24*, 1432. [CrossRef] [PubMed]
26. Cao, J.; Huang, S.; Yu, X.; Li, Y. Acoustic attenuation of side-branched Helmholtz resonator with embedded apertures in grazing flow ducts. *J. Phys. D Appl. Phys.* **2024**, *57*, 205504. [CrossRef]
27. Yaw, Z.; Lai, S.-K.; Gulzari, M. Acoustic resonant metasurfaces with roughened necks for effective low-frequency sound absorption. *Mech. Adv. Mater. Struct.* **2024**, *26*, 2408635. [CrossRef]
28. Lv, C.; Wang, X.; Mei, Y. Optimization Method of Acoustic Filter Structures Composed of Helmholtz Resonators Based on Genetic Algorithm. *J. Phys. Conf. Ser.* **2022**, *2468*, 012019. [CrossRef]
29. Gu, X.; Zhou, G.; Xu, W.; Chen, H.; Qi, C.; Chen, B. Broadband Acoustic Liner Design and Verification of Helmholtz Resonators with Extended Necks. *Noise Vib. Control* **2024**, *44*, 276–281.
30. Jiao, W.; Yang, M.; Ma, Z.; Fu, Y.; Chen, M.; Xie, X. Design and Optimization of 2-MMPPHR Based on Micro-Perforated Panels and Helmholtz Resonators. In Proceedings of the 2024 IEEE International Conference on Mechatronics and Automation (ICMA), Tianjin, China, 4–7 August 2024.
31. Bi, S.; Wang, E.; Shen, X.; Yang, F.; Zhang, X.; Yang, X.; Yin, Q.; Shen, C.; Xu, M.; Wan, J. Enhancement of sound absorption performance of Helmholtz resonators by space division and chamber grouping. *Appl. Acoust.* **2023**, *207*, 109352. [CrossRef]
32. Cai, C.; Mak, C.M. Noise attenuation capacity of a Helmholtz resonator. *Adv. Eng. Softw.* **2018**, *116*, 60–66. [CrossRef]
33. Jena, D.P.; Panigrahi, S.N. Numerically estimating acoustic transmission loss of a reactive muffler with and without mean flow. *Measurement* **2017**, *109*, 168–186. [CrossRef]
34. *ISO 10534-2: 1998*; Acoustics-Determination of Sound Absorption Coefficient and Impedance in Impedance Tubes—Part2: Transfer-Function Method. ISO: Geneva, Switzerland, 1998.
35. *ASTM E2611-12*; Standard Test Method for Impedance and Absorption of Acoustical Materials Using a Tube, Two Microphones and a Digital Frequency Analysis. ASTM: West Conshohocken, PA, USA, 2012.
36. Jena, D.P.; Dandsena, J.; Jayakumari, V.G. Demonstration of effective acoustic properties of different configurations of Helmholtz resonators. *Appl. Acoust.* **2019**, *155*, 371–382. [CrossRef]
37. Wang, L.B.; Ma, C.Z.; Wu, J.H.; Liu, C.R. Realizing high-efficiency low frequency sound absorption of underwater meta-structures by acoustic siphon effect. *Mod. Phys. Lett. B* **2021**, *35*, 2150319. [CrossRef]
38. Long, H.; Cheng, Y.; Liu, X. Reconfigurable sound anomalous absorptions in transparent waveguide with modularized multi-order Helmholtz resonator. *Sci. Rep.* **2018**, *8*, 15678. [CrossRef] [PubMed]
39. Zhang, D.; Tang, W.; Sun, Y.; Chen, C.; Su, X.; Sun, X. Study on Noise-Reduction Mechanism and Structural-Parameter Optimization of Ventilated Acoustic Metamaterial Labyrinth Plate. *Appl. Sci.* **2024**, *14*, 7865. [CrossRef]
40. Faramarzi, A.; Heidarinejad, M.; Mirjalili, S.; Gandomi, A.H. Marine Predators Algorithm: A nature-inspired metaheuristic. *Expert Syst. Appl.* **2020**, *152*, 113377. [CrossRef]

MDPI AG
Grosspeteranlage 5
4052 Basel
Switzerland
Tel.: +41 61 683 77 34

Crystals Editorial Office
E-mail: crystals@mdpi.com
www.mdpi.com/journal/crystals

www.ingramcontent.com/pod-product-compliance
Lightning Source LLC
Chambersburg PA
CBHW041951260926
51850CB00019B/2584
* 9 7 8 3 7 2 5 8 7 2 9 0 9 *